Yuri  London and Manchester

A Smile that Changed the World?

Gurbir Singh

First Published in July 2011
by Astrotalkuk Publications

Copyright © 2011 GurbirSingh

ISBN 978-0-9569337-1-3

Astrotalkuk Publications • Manchester
www.astrotalkuk.org • info@astrotalkuk.org

Cover image courtesy of Ralph Gibson
RIA Novosti Media Library

Rear background image courtesy of Fred Ritchie

To my parents Gurbax and Nanki,
who embarked on their own journey into
the unknown during the heady decade of the 1960s

About the Author

Gurbir Singh is a former college lecturer now working in the IT sector. He has a science and an arts degree. He has been involved in amateur astronomy for over three decades. Currently he is the host of www.astrotalkuk.org, an astronomy blog dealing with topical subjects in astronomy that he established in 2008.

He was one of 13,000 unsuccessful applicants responding to the 1989 advert "Astronaut wanted. No experience necessary" to become the first British astronaut, for which Helen Sharman was selected and flew on the Soviet space station MIR in 1991. Once keen on flying, Gurbir holds a private pilot's licence for the UK, USA and Australia, but does not currently fly.

Born in India, he has been living in the UK since 1966 except one year in Australia. He is married with a three-year-old daughter and lives just outside Manchester in England.

Illustrations

Figure 1 R7 Rocket that launched Vostok (Courtesy Margaret Turnill) 4
Figure 2 The ground track of Gagarin's orbit (Courtesy First Orbit) .. 7
Figure 3 Konstantin Tsiolkovsky's rocket (Courtesy NASA) 10
Figure 4 Konstantin Eduardovich Tsiolkovsky 17th Sept. 1857 – 19th Sept. 1935. (Courtesy NASA) 11
Figure 5 Dr. von Braun and Professor Hermann Oberth being honoured by Berlin Technical University 1963 (Courtesy NASA) 14
Figure 6 Nikolai Kamanin, Boris Belitzky, Yuri Gagarin and Soviet Ambassador Alexander Soldatov at Heathrow Airport, London (Courtesy Patricia Mannarino) 33
Figure 7 Yuri Gagarin with gold medal from the British Interplanetary 40
Figure 8 James Brewster with Yuri Gagarin at the Soviet Embassy ... 42
Figure 9 Yuri Gagarin in co-pilot's seat during flight to Manchester . 46
Figure 10 Hangar #3 at Manchester Ringway Airport 47
Figure 11 Yuri Gagarin arriving at AUFW Headquarters 63
Figure 12 Fred Hollingsworth presenting the AUFW gold medal 64
Figure 13 Yuri Gagarin arriving at Metropolitan Vickers 72
Figure 14 Yuri Gagarin inside Metropolitan Vickers (Courtesy Alf Lloyd) 75
Figure 15 The "Big House" (courtesy Stanley Nelson) 78
Figure 16 Yuri Gagarin at the Cenotaph, London 80
Figure 17 Yuri Gagarin arriving at Manchester Town Hall 82
Figure 18 Yuri Gagarin at Manchester Town Hall (Courtesy Alf Lloyd) 88
Figure 19 Yuri Gagarin on Princess Street returning to the airport 89
Figure 20 Yuri Gagarin leaving Manchester. Upper Brook Street 90
Figure 21 Outside the Tower of London (Courtesy Marx Memorial Library) 95
Figure 22 Yuri Gagarin in London (Courtesy Marx Memorial Library) 98
Figure 23 Captain Eric Melrose "Winkle" Brown (Courtesy Captain Brown) 103
Figure 24 Yuri Gagarin with the Prime Minister for the second time on 13th July 1961 (Courtesy RIA Novosti) 109

Figure 25 Valentina Tereshkova, Secretary-General U Thant, ... 140
Figure 26 Yuri Gagarin and Albert Knight, Moscow, April 1963 144
Figure 27 Painting by Walter Kershaw to commemorate the 50th anniversary 145

Introduction

This book is my attempt to fill in a small piece of the Yuri Gagarin story – his five-day visit to London and Manchester during the summer of 1961. A great deal of excellent work has already been written about Gagarin and the early days of the American and Soviet space programme by authors including Rex Hall, Sven Grahn, David Shayler, James Oberg, Asif Siddiqi, Jamie Doran & Piers Bizony, James Harford and Colin Burgess. Many scientists, engineers, journalists and cosmonauts & astronauts have now also contributed to that story. In writing this book, I have drawn liberally on the meticulous work of these and other authors, some of which are included in the References and Notes section.

I would also like to express my gratitude for encouraging emails from authors who have published many books as I embarked on my first, particularly Michael Cassutt, Asif Siddiqi, Andrew Jenks, Tim Radford and Piers Bizony.

In the first chapter, I provide an overview of the development of rocket technology and the pioneers who conducted the early work and a summary of Gagarin's historic flight. In the final chapter, I reflect on Gagarin's legacy and his relentless calls for peace against the backdrop of increasing tensions of the Cold War, nuclear proliferation, militarisation in East and West Berlin and the first tentative meeting between President Kennedy and Soviet leader Khrushchev in Geneva in June of 1961. Gagarin would have been aware more than most of how unsuccessful those talks had been, which drove his personal quest for peace during his visit to the UK a month later.

In chapter two I cover the troublesome issue for Her Majesty's Government to formally invite Gagarin to the UK, who should greet him on arrival and how to keep Gagarin away from the numerous "communist front organisations" that existed at the time. The tumultuous and warm welcome in London was such a surprise that before the day was out Gagarin had received invitations to meet the Prime Minister and the Queen. In chapters three and four, I cover Gagarin's deep connections to the working class traditions and his personal desire to meet with the foundry workers of Manchester, a trade that he had studied to the age of twenty-one before enlisting in the Soviet air force. Chapters five and six cover his meetings with the Prime Minster and the Queen along with numerous shorter meetings, including the Mansion House, the Royal Society and the Tower of London. With the exception of a short press conference at Heathrow, Gagarin departed the UK without any engagements on the fifth day. A timeline that summarises Gagarin's five days in Britain is included in the appendix.

Although there is just one name on the cover, this work is the collective effort of many. I want to acknowledge the critical contribution of libraries and librarians, especially Central Manchester Library, Rochdale Library, the People's History Museum in Manchester, Mike Weaver and Lynette Cawthra at the Working Class Movement Library, Pamela Clark at the Royal Archives, John Callow at the Marx Memorial Library, John Cunningham at the Society for Co-operation in Russia, members of Salford Astronomical Society, Manchester Astronomical Society and Colin Philip and Andy Green at the British Interplanetary Society.

I am grateful to Reg Turnill for sharing his experience of Gagarin's first post-flight press conference in Moscow which captures the grim anxiety of the east-west relationship at the height of the Cold War. My thanks are also due to Captain Eric Brown for making me welcome to his home and taking subsequent queries on the telephone. An accomplished former chief test pilot, he was the only individual to have a personal one-to-one interview with Gagarin during his five days in

Britain. Surprisingly, this book is the first time Captain Brown had been asked to recall that interview for publication. I have transcribed part of my interviews with Captain Brown and Reg Turnill in chapters six and eight. Video recordings of these interviews are available on my blog.

Others who saw or met Gagarin include Patricia Mannarino, Fred Ritchie, Ray Smith, Marjorie Rose, Stanley Nelson, Stanislava Sajawicz, John Smith and Fred Garner, who at his own expense and effort captured with his cine camera the very moment that Gagarin first set foot in Manchester. Several individuals responded to my calls for witnesses on my blog, posters and newspapers. It is their collective testimony recorded in these pages that is the central contribution of this book. The youngest, Liam Grundy, was four years old at the time, and Dame Kathleen Ollerenshaw will be ninety-nine in the autumn of 2011.

Several people assisted with helpful comments on early drafts, including Vix Southgate and Ralph Gibson from RIA Novosti who shared his time, wisdom and assisted with photos. Chris Welch and others at Gagarin50, Richard Evans and Chris Riley uncovered and shared new materials and have been instrumental in keeping the spirit of the fiftieth anniversary of Gagarin's spaceflight in the public eye. Nick Forder from the Museum of Science and Industry and Professor Jim Aulich shared resources of the Gagarin exhibition in Manchester they had organised to mark the fortieth anniversary. I am also grateful to Dave and Leslie Wright for their suggestion with contacts and prompt sense in fielding my queries. The AUFW no longer exists, so much of the union story emerged with the assistance from members of UNITE Alf Lloyd, Karen Viquerat and Phil Tepper. Professor John Zarnecki and the Very Reverend Philip Buckler Dean of Lincoln answered my initial queries and provided useful corrections and suggestions. This final version is the product of input from three key individuals. Professor Malcolm Heath was extremely generous with his time, reading the entire manuscript and introducing a sense of academic rigour of which I am incapable. Francis French not only read drafts on multiple occasions and provided focused suggestions but also comprehensive support and encouragement from the early stage through to publication.

My special thanks go to my wife Regine, without whom I would still be writing this book. It was not included in the vows we exchanged, but so dedicated was her support that it may as well have been. In addition to providing corrections and comments, she took on large chunks of my portion of childcare to allow me the time to research and write. She also brought her experience as an author of several books to helping with creating an index and the bibliography for this one. I was able to discover a great deal of information, but tantalising questions remain. For example, I was not able to make contact with Fred Hollingsworth (president of the AUFW) or his family. Gagarin met the Queen but I was unable to uncover any pictures.

Transcribing Russian names of individuals, places and events has been tricky. I have attempted to use the "most popular", for example, Yuri rather than Yury or Yuriy. I have included a short glossary in the appendix that includes some of the popular terminology. Despite all the assistance, there will be errors and gaps that remain, for which I and I alone am responsible. I would welcome an email with comments, corrections or suggestions for filling in some of the gaps.

Gurbir Singh,

June 2011

Contents

About the Author ... iv
Illustrations ... v
Introduction ... vii
1. Space Age ... 1
    Around the World in 108 Minutes ... 3
    Rocket Science ... 7
    Twentieth Century Faust ... 16
2. An Uneasy Invitation ... 19
    Communist Party of Great Britain ... 20
    Soviet Trade ... 22
3. Like Some New Columbus ... 30
    Arrival ... 32
    London Press Conference ... 34
    An Ordinary Mortal ... 38
4. Together Moulding a Better World ... 44
    A little bit of History ... 48
    Manchester and Aviation ... 51
    Working Class Traditions ... 59
    Brother Gagarin ... 65
5. Working Class Cosmonaut ... 70
    Firm Handshakes ... 79
    Jodrell Bank ... 83
6. A Delightful Fellow ... 92
    The Royal Society ... 97
    Test Pilot to Test Pilot ... 102
7. Communism and Royalty ... 111
    A Slice of Good Luck ... 114
    Danger of Push Button War ... 121
8. A Smile that Changed the World? ... 127
    Mission Accomplished ... 129
    Good Humoured Evasion ... 130
    Cold War and Peace ... 136
    Peace Envoy ... 140
Appendix ... 148
    Glossary ... 148
    Gagarin's Radio Message to British Foundry Workers ... 150
    Gagarin in Britain: Timeline ... 152
    Bibliography: Books Cited ... 154
    Index ... 156
    Notes and References ... 159

1. Space Age

On the morning of April 12th 1961 two former construction workers, one a specialist in roof tiles and the other a qualified foundry worker, made history with the world's first manned spaceflight to orbit the Earth. Sergei Pavlovich Korolev, the spacecraft's chief designer, was born in 1906 amidst the perils of the Russian Revolution and civil war. Born in 1934, the world's first cosmonaut Yuri Alekseyevich Gagarin grew up in the shadow of the Second World War and the dangers of German occupation. Both went on to study in local vocational schools before turning to their passion for aviation.

Korolev first experienced the magic of aviation sitting on his grandfather's shoulders, at a fairground show in the town of Zhitomyr in Ukraine[1] in the summer of 1913. A biplane, piloted by an early famous aviator Sergei Utochkin, thrilled the crowd who had paid one rouble for the spectacle, as he took off, flew two km and landed again. By his early 20s, Korolev had designed, built and flown his own glider. He completed his pilot's licence for gliders in 1923 and then his single engine Avro 504K biplane in the following year.[2] Intriguingly, Alliot Verdon Roe who designed and built the Avro 504K, was born in Patriot and established factories in nearby Manchester. Around 9,000 Avro 504K were built between 1913 and 1932 in Manchester and under licence in several countries, so it is possible that Korolev's Avro 504K (#353)[3] was built in Manchester.

The success that Korolev's genius brought to the Soviet space programme was not just because of his outstanding skill as an engineer. Nor was it his incredibly generous decision to absolve the Soviet hierarchy for his false imprisonment and mistreatment. It was his, *"formidable political and management skills, reconciling massive egos among his fellow engineers and*

1. Space Age

conducting successful turf wars in the thickets of the Soviet industrial bureaucracy."[4] He convinced Khrushchev to provide him with a direct telephone line to the Kremlin that probably became his most powerful tool in expediting the Soviet space programme.

With only two years of military flying experience of patrolling the Soviet sky in the Arctic Circle, and with just 265 flying hours under his belt, Gagarin formally expressed his interest in spaceflight. In the autumn of 1959, by when he had trained longer as a foundryman than he had been a commissioned air force pilot, Gagarin wrote a note to his commanding officer Lt. Colonel Babushkin, saying, "*In connection with the expansion of space exploration going on in the U.S.S.R. people may be required for manned space flights. I request you to take note of my own ardent desire, and should the possibility present itself, to send me for special training*".[5] During 1959, pilots were interviewed at air force bases throughout the Soviet Union. Following 3000 interviews, twenty pilots were selected for cosmonaut training at a makeshift site on the outskirts of Moscow where today stands the Yuri Gagarin Cosmonaut Training Centre. Surprisingly, about 30% of the 3000 who were initially chosen for further selection decided not to proceed with the programme, preferring to stay with the air force.[6]

Throughout 1960, the 20 cosmonauts were put through a series of programmes of medical tests and training that included parachute jumping, ability to withstand isolation, vibration, acceleration and pressurisation. In the absence of previous experience of training and testing for spaceflight, the training programme was unnecessarily arduous and excessive, conducted in makeshift facilities by specialists who were learning on the job.

1. Space Age

Around the World in 108 Minutes

In preparation for manned spaceflight, the Soviets launched seven unmanned Vostok spacecraft between 15th May 1960 and 25th March 1961, complete with life support, communications and heat shield. Each was designed to go into Earth orbit completely automatically and be recovered. Two failed, two partially failed and three were regarded as successful. On Wednesday morning, April 12th, Yuri Gagarin sat in his 2.4m diameter spherical spaceship attached to the cylindrical equipment module. Both vehicles were secured inside the nose cone of the 34m high 3m diameter R7 rocket initially constructed as the world's first Intercontinental Ballistic Missile (ICBM).

Poyekhali, the Russian word for "let's go", has become the one-word symbol of Gagarin's historic flight on that bright cold Wednesday morning, although it is not something Gagarin remembered saying at the time until reminded of it later. As Gagarin sat in his Vostok spacecraft completely sealed from the outside world with only the radio link to the engineers nearby, he could see only the inside of his 2.4m diameter spherical spacecraft, but a bright light lit up his helmeted face for the live television.

At 09:07 a.m. local time, two stages of this three-stage rocket fired simultaneously, and Gagarin started on his short trip to Earth orbit from the launch site of Tyura-Tam. Two minutes later, Gagarin was thrown forward as the four strap-on booster rockets, each with four engines, were discarded as their propellant supply ran out.

A moment later he was pushed back in his seat as the central, second stage continued the climb with the now lighter load. Gagarin reported, "*The first stage has finished its work. The G forces and vibrations have eased. The flight is continuing normally*".[7] Less than a minute later, the rocket was now through

1. Space Age

most of the Earth's dense atmosphere, the shroud covering Gagarin's spaceship was released and he saw Earth for the last time before entering space. Gagarin reported *"Shroud separation has happened. Through the window, I see the Earth... The ground is clearly discernible. Rivers and folds in the terrain. They are easily distinguishable. The visibility is good. Everything is so clear... My condition is excellent. I'm continuing the flight. The G forces are slightly increasing... I can tolerate everything well. My condition is excellent. I feel cheerful."*[8]

Figure 1 R7 Rocket that launched Vostok (Courtesy Margaret Turnill)

Five minutes into the flight, the second stage central core also exhausted its supply of propellant, detached and fell away with another violent but momentary change in acceleration as the final stage ignited. Gagarin was seated with his back to the Earth, towards his feet was the main window with a periscope device showing him the Earth, as he ascended he reported, *"The third stage is working. The TV light is working. I'm feeling great. I'm feeling alert. Everything is going well. I can see Earth. I see the horizon in the window. The horizon has shifted towards my legs"*.[9]

Following the third stage separation, less than ten minutes after take-off, Gagarin was in Earth orbit. The rocket, now less than a

1. Space Age

third of its original length, pushed Gagarin's spaceship with the attached equipment module to a speed of 8 km/s into an elliptical Earth orbit averaging 250 km high. He reported, *"Separation from the rocket happened at 9 hours 18 minutes 7 seconds, as planned. I'm feeling well. Conditions in the cabin: pressure is one unit, humidity is 65, temperature 20, the pressure in the compartment is one unit... The feeling of weightlessness I'm handling well. It is pleasant. I'm continuing the flight in orbit"*.[10]

The vibration and acceleration ceased, and even though the engines had stopped, it was not quiet.[11] Space is mostly a vacuum where sound cannot travel, but it is far from a silent place. Early spaceflight was not a serene experience: even trying to sleep in a stationary spacecraft on the surface of the Moon required earplugs.[12] The background noise of the pumps and fans for the life support was a reassurance rather than a hindrance. As well as being the first, this would be the shortest of all the manned orbital spaceflights. Through his radio, Gagarin provided continual information on the flight's progress but his voice carried more than just flight data. It was not only what he said but also how he said it that conveyed reassuring signs in his voice, and the television pictures of his face confirmed that exposure to weightlessness was something that human beings could tolerate.

The fear that weightlessness could lead to loss of consciousness or ability to rationally perform basic operations was for the first time demonstrated to be unfounded as Gagarin's coherent responses and demeanour from Earth orbit confirmed. It was in part against this contingency that Gagarin's flight had been designed to be fully automatic. A special four-digit key to allow a manual override of the "autopilot" was provided in a sealed envelope, but it was never required.

Gagarin shared his first views from space of the thin layer of the

1. Space Age

Earth's atmosphere and the stars he saw as he entered the night side over the Pacific Ocean,

"There is such a beautiful halo. Here is a rainbow. First at the Earth's surface. And then the rainbow transitions down. Very beautiful, everything passed through the right window (Porthole). Stars are visible through the periscope. How the stars are moving! It's a very beautiful spectacle. The flight is continuing in the shadow of the Earth. In the right window, right now, I can see a star. She is passing from left to right in the window. She's going, she's going."[13]

No one in Britain could have seen Gagarin's Vostok fly overhead, its flight path did not cover the British Isles. If it had, his spacecraft would have been easily visible with the naked eye as a bright star drifting slowly against the twilight sky. Spacecraft in low Earth orbit can be seen in the night sky during local twilight. Gagarin's trajectory, however, took him into the sunset whilst he was over the largely unpopulated Pacific Ocean and the sunrise whilst in the southern Atlantic Ocean. No record of any visual sighting has ever been published.

Gagarin initially headed northeast towards the Arctic Circle before heading south to the Pacific Ocean, where half an hour after the early morning launch he flew over that part of the Earth where it was night. Confident that Gagarin's orbit was secure, the Soviets announced the successful spaceflight on Moscow Radio. Another half hour later, now close to the south Atlantic and heading north, he experienced dawn where the sun rose as quickly as it had set half an hour earlier. Twenty minutes later, he approached the west coast of Africa, where the scheduled equipment module separation took place, but a connecting cable did not detach. For an additional ten minutes, like an out of control car towing another, the two spacecraft whirled around each other, separating only once the cable had burnt through.

Gagarin's "final approach" took him over the Mediterranean

1. Space Age

Sea, Turkey, over the Black Sea and into the Soviet Union. At 6 km altitude Gagarin ejected, and both he and his spacecraft landed safely by separate parachutes within the Saratov region, which he recognised from the air because it was where he had originally learned to fly only six years earlier.

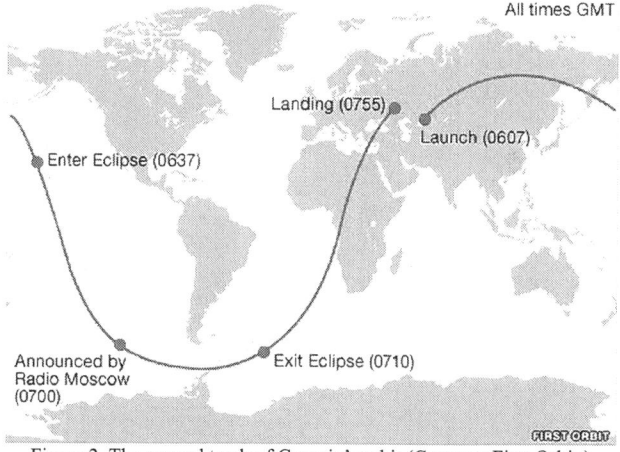

Figure 2 The ground track of Gagarin's orbit (Courtesy First Orbit)

No one outside the USSR knew about Korolev, the chief designer, until after his death in 1967, but Gagarin became a global superstar virtually overnight. Yet his flight was the direct product of the collective work by many other pioneers, most of whom did not live to share that moment on that bright Spring morning of 12th April 1961.

Rocket Science

In the early twentieth century, three men on different continents and in different phases of their careers nurtured the dream of rocket propulsion in pursuit of manned space travel. Konstantin Tsiolkovsky in the Soviet Union, Hermann Oberth in Germany and Robert Goddard in the USA had corresponded with each other but never met. Predominantly products of different nations and cultures, they are perhaps national heroes now, but at the

1. Space Age

time they were unknown even amongst their own citizens. Collectively, they were embarked on a line of research that in a few decades pushed science into territory firmly occupied by science fiction for millennia. Between them, they laid the theoretical groundwork that was transformed into a viable rocket by the efforts of two individuals who are permanently associated with the history of rockets, Wernher von Braun and Sergei Pavlovich Korolev.

Earlier, a Russian like Tsiolkovsky arguably made the first substantial contribution to rocket science under extraordinary personal circumstances in the late nineteenth century. In the last seventeen days of his 27-year-old life, Nikolai Kibalchich, condemned and imprisoned in the fortress of Petropavlovsk in St Petersburg, had hastily sketched a design and his thoughts for a radical new vehicle he called an "aeronautical machine". Two decades before the Wright brothers demonstrated their invention of the aeroplane in 1903, Kibalchich arguably produced the world's first rocket design. The radical element in Kibalchich's aeronautical machine, apart from being a man-made flying machine, was the mechanism of propulsion. He concluded that steam engines and electric motors did not possess the required energy efficiency and were too large and heavy, so proposed a slow burning explosive as an alternative. A *"quintessential bomb-maker cum scientist"*,[14] he was familiar with explosives; he had used them to assassinate the Tsar.

Nikolai Kibalchich, the son of a priest, was arrested for taking part in the assassination of Tsar Alexander II on 1st March 1881. He was found guilty and, with four others, hanged early in the morning of 3rd April 1881. During the two weeks between his arrest on 17th March and his execution, within the confines of a dimly lit prison cell using a simple pen and paper drawing only on what he could remember, he hastily documented a design for his "aeronautical machine".

1. Space Age

Despite Kibalchich's pleading, his paper was not communicated to scientists but filed away by the police two days before his execution and remained unseen for thirty-six years.[15] In August 1917, as Russia entered another perilous phase of its history, this design was rediscovered and published by a historian of aeronautics, Nikolai Rynin. Kibalchich, a rocket scientist and a nineteenth-century terrorist, writing a few days before his execution began his paper with, *"Being in prison, I am writing down this project a few days before my death. I believe that my ideas can be realized, and this belief strengthens me in my terrible situation"*.[16]

Kibalchich visualised compressed gunpowder placed inside a long cylinder, completely sealed with the exception of a nozzle underneath where hot gases from the combustion could escape forcing the cylinder to rise. He called the cylinder a "gunpowder candle". He calculated that by compressing the gunpowder, it would not instantaneously explode but would "combust slowly", and as long as the force generated by the escaping hot gases exceeded that of the weight of the cylinder, the cylinder would ascend. Hovering, he determined, could be achieved by using thinner gunpowder candles, and to lower the aeronautical machine to the ground he proposed the use of progressively decreasing diameter gunpowder candles. A smaller secondary gunpowder candle placed on the outside and movable like a ship's rudder would allow complete directional control. Kibalchich acknowledged the theoretical nature of his work and concluded,

"It will be possible only by means of experiments to prove whether my idea is correct or not. Besides, only tests can establish the necessary relationships between the dimensions of the cylinder, the diameter of the gunpowder candle, and the weight of the machine to be lifted. Initial experiments may be conveniently carried out with small cylinders even in a room."[17]

1. Space Age

Given his mortal predicament, Kibalchich worked astonishingly diligently and, at a time before aeroplanes, he was remarkably creative. His propulsion system, then known as reactive engines, was impractical, and had he lived longer maybe he would have refined these calculations through experiment to propose a more viable aeronautical machine. His vision was limited to his aeronautical machine travelling only within the atmosphere. He makes no mention of space. The concept of rockets beyond the Earth's atmosphere, of space travel, required an additional leap of imagination that came from a fellow Russian, Konstantin Tsiolkovsky, in a landmark paper published in 1903, the same year that the Wright brothers demonstrated the first flight of a powered, heavier-than-air manned aircraft.

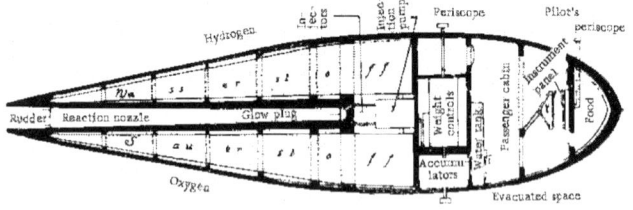

Figure 3 Konstantin Tsiolkovsky's rocket (Courtesy NASA)

This partially deaf, largely self-taught teacher of mathematics living in a provincial town on the Western edge of the Soviet Union solved quantitatively many of the fundamental problems associated with reaction engines in 1903.[18] Konstantin Tsiolkovsky is considered as the father of rocketry because he derived the key concepts of space travel and developed a mathematical formula for calculating rocket speed, identifying key benefits of liquid rather than solid fuel and the idea of staging, where a rocket discards a segment during flight to reduce weight. He correctly visualised and articulated concepts of acceleration and weightlessness, and formulated his ideas through the rigour of mathematics. He calculated and published for the first time the speed of 8 km/s necessary for a spacecraft to attain Earth orbit and 12 km/s for a spacecraft to overcome Earth's gravitational pull necessary for interplanetary travel.[19]

1. Space Age

Tsiolkovsky dedicated most of his 78 years to rocket technology, often through experiment but mostly through mathematical analysis. He laid the groundwork for the space age, which emerged after the Second World War. He designed the smooth aerodynamic shape of rockets, tanks for liquid fuel, a rocket nozzle for combustion and pressurised living quarters.

Figure 4 Konstantin Eduardovich Tsiolkovsky 17th Sept. 1857 – 19th Sept. 1935. (Courtesy NASA)

Gagarin's spacecraft Vostok had three portals, one with a periscope that looks as if it was inspired directly by Tsiolkovsky's design. At a time before air travel, Tsiolkovsky describes vividly and with astonishing accuracy the experience and danger of a rocket launch, *"The signal has been given; the*

1. Space Age

explosion, accompanied by deafening noise, has started. The rocket has shivered and is on its way".[20]

Tsiolkovsky would not have experienced travelling at a speed faster than that of a train but he describes the acceleration and associated G-forces with dramatic clarity, *"We experience a tremendous increase in weight. The 4 [units] of my weight have been transformed into 40 [units]. I have fallen to the floor, have been knocked out or even have died."*[21] The epitome of Tsiolkovsky's genius was to imagine a sensation for which he could have had no preparation. The spectacle of weightlessness was a sensation still alien to humanity when he wrote:

"The hellish gravity which we experience will last for 113 seconds or approximately 2 minutes, until the explosion and its noise have ceased. Then when dead silence occurs, gravity disappears as quickly as it appeared. Now we have risen beyond the limits of the atmosphere to an altitude of 575 km. Gravity has not only weakened: it has disappeared without any trace... All the objects which have not been secured inside the rocket have moved from their places and hang in air, not touching anything; and even if they are touching, they do not exert pressure on each other or on the supports... we float in the middle of the rocket like fish."[22]

Tsiolkovsky's extensive work motivated many of the key individuals including Sergei Korolev on the road to innovative technological developments responsible for the swift achievements of the early Soviet space programme. Speaking on BBC Radio News on the day after Sputnik was launched, Bernard Lovell commented, *"this success has been achieved by a nation which a generation ago was largely illiterate".*

At 09:32 a.m. on 16th July 1969, Apollo 11 started humanity's greatest adventure, the first manned landing on the surface of the Moon. Amongst the VIPs at the press stand at the Kennedy

1. Space Age

Space Centre were Arthur C. Clarke, Charles Lindbergh and a 75 year old Hermann Oberth who had personally corresponded with Tsiolkovsky almost half a century earlier. Immediately after the launch, Oberth said via an interpreter *"When I started thinking of this flight I was a boy of 11. It was just as I imagined, only more marvellous"*. [23]

Hermann Oberth, a German born in modern-day Romania lived in various countries in Europe including Germany, Austria, Hungary, Romania and Italy during one of Europe's most tumultuous centuries. Space historians consider Oberth as one of the founding fathers of rocketry. He contributed to its development through theory and experiment. His examiners at the University of Heidelberg rejected his 1922 doctoral thesis on rockets and space travel, but he published it privately, and in 1923, he wrote a book that became a classic: *The Rocket into Interplanetary Space*. A young student, Wernher von Braun, fascinated by rocketry, bought this book, which resulted in two significant decisions that shaped his destiny. First, it led him to focus on maths and physics as subjects to study[24] and later in Berlin when Oberth began testing liquid fuel rockets, von Braun began to work for him on Oberth's request.

Oberth is credited for independently coming up with the principles of rocket staging, the use of liquid fuel for propulsion, and in 1923 the idea of manned spaceflight using rockets. Tsiolkovsky disputed this claim, insisting that Oberth's work was being accepted as a new discovery whereas he had already proposed the same ideas back in 1903.[25] Three decades older than Oberth, Tsiolkovsky was one of the earliest champions of rocketry. In 1929, Oberth expanded his original book to a larger 429-page book that he called *Ways to Spaceflight*. He sent a copy to Tsiolkovsky, asking for a copy of his latest work in exchange.[26]

The combustion of liquid fuel can be regulated, increased,

1. Space Age

decreased, stopped and restarted. Any engines designed with solid fuel, however, "gunpowder candles" for example as envisaged by Kibalchich, are inherently uncontrollable once ignited. During the period between the wars, liquid fuel propulsion for rockets became the focus for research. Germany was prevented from researching new aircraft by the Treaty of Versailles (signed on 28th June 1919 ending the state of war between Germany and the allied powers), but the treaty had no restrictions on offensive missile technology because it did not exist when the treaty was signed.

Figure 5 Dr. von Braun and Professor Hermann Oberth being honoured by Berlin Technical University 1963 (Courtesy NASA)

In 1934, a special rocket technology unit headed by artillery captain Walter Dornberger was established by the German army.[27] Dornberger immediately recruited a twenty-two-year-old Wernher von Braun who had completed his PhD thesis that year[28] in which he investigated the use of liquid fuels for rocket propulsion.

The first successful launch of a rocket using liquid propellant (liquid oxygen and gasoline) had already been conducted by

1. Space Age

Robert H Goddard on a family owned farm in Massachusetts on 16th March 1926. A small rocket of 7.25kg, most of which was the fuel, travelling at almost 100 km/h, took off and landed 20 seconds later 55 metres away. The statistics, like those of the first powered flight by the Wright brothers, are not impressive; this practical demonstration, this proof that liquid fuel could be used as means of rocket propulsion, was however profound. None of Tsiolkovsky's or Oberth's experiments had yielded such critical results.

Despite his pioneering success, Goddard did not get the support from the US military even though he sought it on several occasions. In contrast to his contemporaries, particularly Wernher von Braun and Sergei Korolev, it was Goddard's *"reticence to seek publicity"*[29] that initially prevented the international and national recognition that his achievements deserved.

Goddard continued to enhance his techniques following his success in 1926 and filed 48 patents during his lifetime; another 35 had been submitted but not processed when he died in 1954. His wife who had supported and filmed much of his early work as the executor of his estate submitted a further 131 patents from his sketches and notes. In 1960, the recently formed National Aeronautical and Space Administration acquired the rights to his patents for a fee of one million US dollars.

In 1920, Oberth wrote to Goddard requesting a copy of his book *A Method of Reaching Extreme Altitudes*. In the following year, Oberth reciprocated and sent Goddard a copy of *By Rocket into Planetary Space*, where he acknowledged Goddard's work and asserted that his theoretical work was supplemented by Goddard's experiments. Despite this, Goddard, like Tsiolkovsky, was convinced that Oberth was "borrowing" his ideas.[30]

How was it that these early rocket scientists, in different

continents and working in various languages, communicated so readily? Tsiolkovsky fostered cooperation by publishing the names and addresses of interested individuals on the back pages of some of his publications.[31] This 1920's answer to Facebook facilitated international collaboration, including occasional abrasive exchanges. In 2007, the young but prolific space historian Asif Siddiqi uncovered letters[32] that Goddard sent in 1924 to the USSR. That seed, the earliest cooperation in space exploration between USA and USSR, inspired another. Three hundred kilometres above the French city of Metz,[33] at 2:17 p.m. on Thursday 17th July 1975, a Soviet and an American spacecraft docked and the commanders of the Apollo-Soyuz Test Project Alexi Leonov and Tom Stafford shook hands in space.

Twentieth Century Faust

Fifteen years almost to the day before the launch of the world's first artificial satellite, Wernher von Braun, accompanied by his mentor Hermann Oberth, celebrated with a banquet in Schwabes Hotel in Peenemunde on the north German coast the first successful launch of the A4 rocket, later renamed as the V2.[34] The V stands for "Vergeltung", which is frequently mistranslated as "vengeance" but is more accurately translated as "reprisal". There had been many failures before and there would be more to come. With many aspects of the new and complex engineering (control surfaces, gyroscope, and propellant line valves) still to refine, but on the night of 3rd October 1942 they celebrated their technological success in building the world's first supersonic weapon of mass destruction.

This success did not go unnoticed, and on 17th August 1943, a RAF raid by almost 600 Halifax and Lancaster bombers on Peenemunde killed 735 German soldiers and rocket engineers at the cost of 300 RAF personnel in 40 aircraft shot down during the return flight. It was very fortunate for one German, "*it saved my life*" recalled Hans Endert in 1990 during an interview with

1. Space Age

the BBC aerospace correspondent.[35] Endert had been serving as a telecom engineer on the Russian front and he was called back to Peenemunde to help replace the engineers lost during the RAF raid.

Reg Turnill had spent most of the 1960s between Houston, Texas and Florida, covering the American space programme for the BBC. Through numerous TV interviews, he had got to know von Braun well, but being familiar with his past did not shake hands for the first two years. One of the first V2 in 1944 came down close to Turnill's south London home and hastened the arrival of his first son.

In 1993, Turnill visited the Dora-Mittelbau factory north of Nordhausen in the mountain tunnels where the V2 production was moved to in October 1944 fearing another RAF raid. Of the 60,000 concentration camp workers forced to work on building the V2, 20,000 died in the camp, far fewer than those killed by the 3000 V2 launched before the war ended. In the hasty retreat from Dora-Mittelbau as the Red Army approached in 1945, von Braun was quoted saying, "*We fired missiles against England, but it's the Russians who will take revenge*".[36] Given that a majority of the slave workers were from the countries that would eventually form the Eastern Bloc, including a large proportion of Russians, his assessment was probably correct.

A standard military tactic requires that a retreating army destroy assets to ensure they do not fall into enemy hands. The documents the von Braun team had produced, von Braun himself and his team were such an asset. Having escaped from the Red Army, once the German army degenerated, von Braun had to escape from the Gestapo, too. He along with his team of engineers survived and surrendered to the American forces with his research intact.

Violence, cruelty and suffering were an integral part of the story

1. Space Age

of rocket development. Korolev had been a victim and suffered whilst von Braun had sided with the perpetrators, each driven by his instinctive, singular and deep-seated desire for space travel. With his first-hand experience of von Braun, Turnill had concluded early on that von Braun was a "Twentieth Century Faust" having sold his soul in exchange for an opportunity to fulfil his personal ambition to build rockets that could reach the Moon.

The "space race" that ensued in the 1960s had no clear start or end. The Soviets took the first specimens of life from Earth (including turtles, flies and worms) in September of 1968 to the Moon, three months before three astronauts of Apollo 8 arrived there in December. If any one instance marks the point the Soviets lost then it was a meeting in April 1969 between Vasily Mishin (who replaced Korolev in 1966) and the Communist Party General Secretary Leonid Brezhnev. During that meeting, Mishin explained why the Soviet space programme was falling behind and uttered words which in another setting would have been considered sacrilege. He questioned the fabric of the Soviet system by suggesting that they were behind the Americans because "*the lack of material incentives among workers*".[37]

2. An Uneasy Invitation

In July 1961, a 27-year-old Soviet air force Major, Yuri Alekseyevich Gagarin, arrived in the UK. He was still the only person to have been in Earth orbit. He had flown higher (327 km) and faster (27400 km/h) than anyone before him. He saw the Earth in its entirety, experienced strong, hot sunlight undiluted by the Earth's atmosphere, glimpsed with his naked eye the brightness of stars against the blackest sky whilst subjected to an extended sensation of weightlessness. For many who saw him in London and Manchester, this was a rare encounter. He was a representative of humanity testing a completely new environment from which he could not have been certain of a safe return.

The uneasy friendship between the allies (Britain, Soviet Union, France and the United States) during World War Two evaporated into the Cold War, in some respects even before the War had ended. The Berlin blockade between 1948 and 1949 was the first major dispute, and another was brewing at the time of Gagarin's first visit to the west in 1961. The Soviets in pursuit of their political ideology built a wall around West Berlin, initiating a political stalemate that would endure for almost three decades.

Neil Armstrong, Buzz Aldrin and Michael Collins' world tour after their momentous landing on the Moon in July 1969 was to follow in the footsteps of Gagarin's hectic world tour following his historic space flight on 12th April 1961. Between late April and early August 1961, Gagarin visited ten countries, and a week after his UK visit, Gagarin was in Poland and Cuba. When Gherman Titov took off on Vostok 2 (the second Soviet manned space flight) on August 6th 1961, Gagarin was in Canada and headed back to the USSR to be at Red Square for the celebrations on Titov's return.[1] His first trip, at the request of the Czechoslovakian president, was to Prague on 28th April 1961.[2]

2. An Uneasy Invitation

That invitation had come from the head of state, but a formal invitation from the UK was problematic in several respects. The Soviets would exploit the visit purely as a propaganda opportunity, it would be incongruous given their almost belligerent behaviour over Berlin, and it would be embarrassing for Britain's American allies.

Communist Party of Great Britain

During the 1960s Communism had a stronger presence within British politics than it has today, and like all other protagonists it intended to exploit the cosmonaut's visit for its own interests. The Communist Party of Great Britain (CPGB) had been founded in 1920 by the merger of several organisations considered as socialist, Marxist or left-wing. Individuals from the Labour Party, members of the "Hands off Russia" campaign, shop stewards and workers' committees and even an elected MP, Cecil L'Estrange Malone from the British Socialist Party, came together to form the CPGB. In January 1921, the CPGB was refounded after the memberships of Suffragist Sylvia Pankhurst's group and the Scottish Communist Labour Party and others agreed to join.

Declassified papers from the Foreign Office (FO), today is known as the Foreign and Commonwealth Office (FCO), reveal the political considerations behind the reluctance by Her Majesty's Government (HMG) to extend Gagarin a formal invitation. When, or even if, he would actually come to Britain remained uncertain as both governments attempted to secure their own political interests. The declassified papers reveal for the first time the ad-hoc, last-minute arrangements that shaped the cosmonaut's first visit to Western Europe. During a visit to London on June 30th, Boris Borisov, the Soviet deputy leader for Foreign Trade, stated that Gagarin will "*probably be at the Trade Fair in London in July*".[3] Only a week before his arrival, Soviet officials were still denying that Gagarin would visit the UK,[4] and

2. An Uneasy Invitation

then suddenly on July 7th, the day the Prime Minister opened the Soviet Trade Fair at Earl's Court, it was announced that a diplomatic visa for Gagarin to visit the UK had been granted. This was the first indication to the Amalgamated Union of Foundry Workers that the invitation that Gagarin had accepted in May to visit them in their union headquarters in Manchester would actually happen.

Once an announcement was made that his UK visit would proceed, additional invitations from within the UK came from industry, trade unions, businesses, and engineering and scientific institutions. A diplomatic confirmation recorded in Foreign Office memo came from the Soviet Embassy. He would arrive on July 11th with four companions and stay for *"two or three days"*.[5]

The Soviet government juggled with the prioritisation, propaganda and sheer logistics, but ultimately there were many more invitations than could have been practically accepted for the envisaged length of the visit. Initially, the confirmation of Gagarin's intention to visit the UK stated that he planned to stay for three days.[6] During the four days between the issuing of his visa and his arrival at Heathrow he received an invitation from the Prime Minister and so his visit was extended to four days. On his first day in the UK, the 11th July, when he was attending the Trade Fair in Earl's Court, Gagarin received a formal invitation to Buckingham Palace for Friday 14th. So, yet another day was added to his stay in the UK, extending it to five days. These extensions enabled him to accept additional invitations, but amongst those he was unable to accommodate was one to Jodrell Bank. He would have personally liked to have visited the radio telescope that had contributed to the Soviet space programme but his itinerary could not accommodate it.[7]

Accordingly, during his five days in Britain the itinerary evolved with many of the visits being offered, changed, and agreed at

2. An Uneasy Invitation

very short notice. He started and ended each day of his visit at the Soviet Embassy in Kensington.

Soviet Trade

The Soviets' desire to link Gagarin's visit with the Soviet Trade Fair in London in Earl's Court between 7th and 29th July 1961 limited the dates Gagarin could choose for the visit to Britain. Summer 1961 was a time of global political and military unrest, one of the colder phases of the Cold War. Three months earlier the Americans' failed invasion of Cuba at the Bay of Pigs had undermined the credibility of the recently elected President Kennedy. Three months later a disagreement over West Berlin within the erstwhile allies had led to the building of the Berlin Wall. The mutual ban on nuclear testing agreed between the USA and the USSR in October 1958 ended on September 5th 1961 when first the USSR and then the USA resumed nuclear testing. The number and yield of bombs tested introduced a potential for destruction the world never before experienced. These circumstances conspired to set in motion the events leading to one of the most dangerous stand-offs in the twentieth century, the Cuban Missile Crisis in late 1962. At the time of Gagarin's visit, the severe east-west political anxiety and the high likelihood of military action was however not fully appreciated by a wider public, as this, now declassified, Foreign Office memo warned:

"If, as I think, the Soviet Government are anxious that Major Gagarin should receive an official invitation to visit the U.K., we should consider what their motive might be. They are well aware that he would be given a hero's welcome and would be feted up and down the country. The first man into space and the personification of Soviet achievements in the field of rockets and ballistic missiles is a young man with a frank and engaging personality. The Soviet Government might well calculate that his appearance in this country at a moment when the issue of

2. An Uneasy Invitation

Germany and Berlin is moving steadily towards a crisis and when HMG. and allied governments are seeking an answer to the question of whether they are prepared in the last resort to face nuclear war over Berlin, would incline public opinion in this country towards accepting the view that the Soviet case in regard to Germany and Berlin is reasonable and that although the Soviet Government clearly have the capacity to launch nuclear devastation, they have no intention of doing so."[8]

In the same memo R. H. Mason (head of the Northern department) goes on to report that he had been informed by Mr Yarotsky, Counsellor of the Soviet Embassy, that "*Gagarin had received invitations from over fifty individuals and organisations in this country including the mayor of Newcastle, several trades unions and British Interplanetary Society.*"

Consequently, by June 1961, although willing to welcome him, HMG was unwilling to offer a formal invitation for Gagarin to visit the UK. An invitation is exactly what the Soviets were seeking. An official invitation from Britain would highlight the relatively weak achievements in space by the US, although astronaut Alan Shepard had by then successfully completed America's first sub-orbital spaceflight. A report in the Soviet newspaper *Izvestia* on 19th June 1961 would potentially resolve this quandary. In that report, Sir Fitzroy MacLean, president of the GB-USSR Association, had sent an open invitation for Gagarin to visit the UK whenever he liked and stay as long as he liked. The report published details of this invitation, which had been made in conjunction with two other societies associated with Russia - the British Soviet Friendship Society (BSFS) and the Society for Cultural Relations between the British Commonwealth and the USSR (frequently abbreviated to SCR). The BSFS and SCR were considered merely as "*communist front organisations*"[9] with dubious affiliations and were cast as "strange bedfellows"[10] or "undesirable groups"[11] in the Foreign Office papers. In contrast, the GB-USSR

2. An Uneasy Invitation

Association was a HMG sponsored organisation, and although it had made the invitation prior to consultation with HMG, this indirect invitation obviated the requirement for HMG to instigate an official HMG invitation.

The UK Commonwealth and Foreign Office had funded the GB-USSR Association with a remit to foster contact and dialogue between GB and the then countries of the Soviet Union, which thus formed an ideal group to invite Gagarin under apparently non-political auspices. At its peak, it had a membership of around 1500 members, many of whom were senior politicians and academics. Between 1970 and 1977, the former Prime Minister Harold Macmillan would serve as its president. Following the break-up of the Soviet Union, it was reformed as a non-governmental organisation called The Britain-Russia & British East-West Centres. It restructured again in 2002 with a more manageable name as the GB-Russia Society.

Another invitation had come from an unusual source, and on equally unusual reasons would ultimately fail. Victor Hochhouser and his wife Lillian had been (and still are) involved with promoting artists from around the world to visit the UK. From the early 1960s onwards, their impressive lists include the Bolshoi and Mariinsky (formerly Kirov) opera and ballet companies, musicians including the Borodin Quartet and Emil Gilels. They also supported major international companies such as the Guangdong Acrobatic Troupe of China, Paco Peña and his Flamenco Dance Company, the National Ballet of China, San Francisco Ballet and the Spanish Riding School of Vienna. This could have been a non-political platform for Gagarin's invitation, but ultimately politics intervened.

Whilst in Leningrad in April 1961, organising the Leningrad Kirov Ballet for a four-week season in Covent Garden, Hochhouser had invited Gagarin to the UK for the opening performance in June. Convinced by the positive response he

2. An Uneasy Invitation

went public with his claim that *"Gagarin may come to London on "first night" of the Leningrad Kirov Ballet performance"*.[12]

On June 17[th,] the Leningrad Kirov Ballet troupe were leaving Le Bourget Airport at the end of a three-week tour in France to start a four-week season in London. The principal dancer, Rudolf Nureyev, however, chose that moment to seek political asylum. The Soviets did not want Gagarin to be associated with a high profile defection. The Leningrad Kirov Ballet troupe continued their journey to London without their principal dancer, and Hochhouser's efforts to bring the cosmonaut to Britain were wrecked.

Hochhouser's press coverage had led HMG to pre-empt the potentially awkward situation of how the UK might honour Major Gagarin if he were to visit the UK.[13] The Under Secretary for the Foreign Office Sir Fredrick Hoyer Millar confidentially wrote on 21st April 1961 to the Royal Institution of Chartered Surveyors (RICS) encouraging them that they, as a private organisation, should consider honouring Gagarin.[14] The RICS had honoured explorers like Edmund Hillary who climbed Everest 1953 and Vivian Fuchs who made the first overland crossing of Antarctica in 1958. By keeping Gagarin's diary full with invitations from organisations of which HMG approved, there was less likelihood of Gagarin accepting invitations from "less desirable" groups. A record of the RICS' response is not available in the archives and there is no evidence of Gagarin actually visiting the institute during his five days in the UK.

A British Trade exhibition in Moscow in May 1961, where a meeting between Khrushchev and HMG representatives was inevitable, would have been an ideal moment to extend a formal HMG invitation for Gagarin to visit the UK. The British Trade and Industrial Exhibition in Moscow included a variety of British technical and medical instruments, toys with movable limbs, the latest British fashion, a model of the new BOAC

2. An Uneasy Invitation

airliner and a model hovercraft. The Soviet leader Nikita Khrushchev visited the British Trade and Industrial Exhibition in Moscow accompanied by Reginald Maudling. At the Associated Electrical Industries (AEI) Limited stand Khrushchev met Gilbert Jolly, a Russian-speaking director of exports at AEI Metropolitan Vickers. Jolly would in due course personally welcome Gagarin during his visit to AEI Metropolitan Vickers in Trafford Park during an overcast lunchtime on Wednesday 12th July.

The question of HMG extending such a formal invitation for Gagarin to visit the UK was addressed within a parliamentary question on 27th April from Norman Pentland MP for Chester-Le-Street. He broached the topic by asking "*the President of the Board of Trade whether, on the occasion of his visit to the British Trade Fair in Moscow next month he will extend an official invitation to Major Yuri Gagarin to visit Great Britain*".[15] The president of the Board of Trade responded with "*I understand that an invitation is being offered. I do not think we should butt in, but certainly everybody will be delighted if such an invitation is extended and accepted*", skilfully steering away from commenting on the awkward issue of an official government invitation.[16]

Throughout May and June, political and diplomatic, mostly unofficial, exchanges referenced indirectly in the Foreign Office papers illustrate the Soviets' eagerness for an official invitation and HMG's reluctance to offer one.

On 7th July, at the start of the Soviet Trade Fair, the Soviets blinked and announced that Gagarin would come to the UK, formally responding to the invitation from the organisers of the Trade Fair. Gagarin officially accepted the invitation from the managing director of Industrial Trade Fairs limited, Mr V.G. Sherren, the company that had organised the Soviet Trade Fair in London. Despite the absence of a formal HMG invitation, the

2. An Uneasy Invitation

Soviets had perhaps concluded that the propaganda value was substantial enough for the visit to proceed. A formal invitation from HMG was never made.

Throughout his visit, Gagarin would be accompanied by Lieutenant-General Nikolai Kamanin who as the military chief was the head of the cosmonaut training programme, and who would chaperone his new global superstar and provide the necessary security and political guidance. As Gagarin did not speak English, his translator for the duration of his visit was Boris Belitzky. Other members of his team included Nicolay Denisov, his biographer, and since this was an official UK visit, the Soviet ambassador Alexander Soldatov would be present for formal events.

A notable absence from Gagarin's entourage was his wife. On March 7[th,] 1961, Gagarin's wife Valentina had given birth to their second daughter. On later trips Valentina would travel with him, but in July she was still at home nursing their five month old daughter Galina.

Lieutenant-General Nikolai Kamanin was a Soviet pilot hero in his own right. He was one of seven pilots who had rescued the one hundred and eleven crew members of the steamship SS Chelyuskin during a hazardous expedition to determine the possibility of travelling by non-icebreaker through the Northern Maritime Route. Although reinforced, SS Chelyuskin was not an icebreaker, and its perilous journey ended when the ship succumbed to the ice fields in September 1933 and drifted for months. Eventually crushed by the icepacks near Kolyuchin Island, it sank. The crew escaped onto the ice and survived in makeshift shelters through the harsh arctic winter. In February 1934, a series of aircraft landed on a primitive landing strip which the crew made with shovels and had to remake several times. Seven pilots using three different aircraft rescued all one hundred and eleven crew; Kamanin was one of the seven pilots

2. An Uneasy Invitation

flying a two-seater Polycarp R-5 biplane.

For this he was awarded the title "Hero of the Soviet Union", an award established in April 1934, just in time for these pilots. "Hero of the Soviet Union" was awarded for outstanding feats of distinction in the service of the Soviet state personally or collectively within a group. By the time the USSR ceased to exist on 31st December 1991, 12,745 awards of "Hero of the Soviet Union" had been awarded. Kamanin's was the sixth. Following his spaceflight Gagarin was also awarded one: his was number 11,175. All flown cosmonauts during the early days of the Soviet space programme received one too. Kamanin went on to win further awards for his combat missions during World War Two in the Soviet Union, Hungary and Romania. His son Arkady at the age of fifteen was the youngest pilot serving in the Soviet air force during the Second World War, winning the Order of the Red Star and Order of the Red Banner, but he died in 1947.

As the Director of Cosmonaut Training, Kamanin had been instrumental in selecting Gagarin as the first cosmonaut and would remain in charge through all the triumphs and tragedies of the cosmonaut programme until 1972. He was a man of undisputed courage, a strict disciplinarian and an effective manager with a complex personality. He was described as *"an aging war hero"*, an *"authoritarian space tsar"*[17] and a *"complete Stalinist bastard"*.[18] He would accompany Gagarin to virtually all of his engagements including those in Manchester, the meeting with the Prime Minister and to Buckingham Palace.

Boris Belitzky, a well respected and experienced member of the English department of Radio Moscow, was in London already to cover the Soviet Trade Fair at Earl's Court, which officially had opened on 7th July for three weeks. It was after Belitzky had arrived and started his preparation for the coverage of the trade fair that he received instructions from the Soviet Embassy in

2. An Uneasy Invitation

London that he would be Gagarin's interpreter throughout Gagarin's visit in Great Britain.[19] Belitzky as the sole translator accompanied Gagarin through his UK visit and turned his attention to the Soviet Trade Fair only after Gagarin had returned to Moscow.

Belitzky was an experienced journalist and presented a popular programme on radio called "Science and Engineering" on *Voice of Russia*. Not only did he have an excellent command of English but he also had an exquisite voice, which he developed over fifty years as a journalist. Without a hint of an accent, he could easily have been mistaken for an announcer on the BBC. The previous year he had been asked to translate for the high profile trial of Gary Powers, a US air force pilot of a secret high altitude surveillance aircraft that had come down in USSR territory during a high-altitude spying mission. Powers' trial in Moscow was a particularly embarrassing episode for the US President Eisenhower and Vice President Nixon, who was quoted to be having a *"nervous fit"*[20] because US public opinion considered US foreign policy to be on trial rather than the captured pilot.

Belitzky would be present at all the receptions, including the one at Buckingham Palace. Despite being surrounded by food and drinks at the numerous daily lunches, receptions and dinners throughout the five days, he as the translator would not always have the opportunity to eat and would occasionally go hungry. The ad-hoc, sometimes chaotic, decisions in the end resulted in an entourage comprising a team of Soviets that would successfully extol the values of the Communist party in the heart of Europe.

3. Like Some New Columbus

The idea of space is ubiquitous today; fifty years ago, it was just emerging from the realm of science fiction. It is difficult to appreciate achievements in human spaceflight for those born and living in the space age. Life in the twenty-first century is routinely touched by space age technology through satellite television, GPS enabled mobile phones, daily weather forecasts and public outreach by those who have been in space. Information in databases populated directly by orbiting satellites is available with unimaginable speed with an internet connection from home or a mobile phone. *"Any sufficiently advanced technology is indistinguishable from magic",*[1] wrote Arthur C. Clarke. The crowds that lined the roads in London that day to see the cosmonaut saw someone who may as well have come from another world. He was a real life manifestation of a comic book superhero, a character from science fiction, someone who came out of nowhere and stepped into their world as if by magic.

Gagarin's first day in Britain on the face of it looked uncomplicated. Following his arrival at the airport he was driven the twenty-two kilometres to the Soviet Embassy in Kensington. From there it was a short hop to Earl's Court for the Soviet Trade Fair, where lunch and a news conference were scheduled along with a tour of the exhibition. The first day in the UK would conclude with a formal reception at the Soviet Embassy. The Soviet Embassy served as his home, and each of his five days in Britain would start and end from this base.

Socialist organizations and left-wing media were already prepared to exploit the Soviet Trade Fair to advance their socialist agenda, and now with Gagarin's presence for the first time in Western Europe there was a prospect of an unsurpassed propaganda opportunity. Gagarin's good looks, easy going personality and friendly nature made him a natural and ideal icon of Communism. Most of all, he had achieved something truly heroic. No one could doubt his courage. Despite the

3. Like Some New Columbus

elaborate testing, at the time of his launch there was no guarantee he would safely return to Earth. Some estimates gave him a fifty-fifty chance of success. Prior to his flight, Gagarin expressed serious doubt in his mission. In a recent interview, his daughter Elena recalls a letter that Gagarin wrote to his wife prior to the mission. Elena recalls the content, *"it was likely he wouldn't return, because the flight was extremely dangerous, and that he wanted her not to remain on her own in that case"*.[2]

The Communist Party of Great Britain (CPGB), the Anglo-Soviet Friendship Society, the Society for Cultural Relations and some elements of the Labour party would exploit Gagarin's success and presence in the UK to assert their own left wing working class ideals.

Gagarin became the object of media scrutiny wherever he went. The communist broadsheet, the *Daily Worker*, covered his visits with more gusto than others. By the early 1960s, the *Daily Worker*, originally a CPGB broadsheet founded in 1930, was the only longstanding socialist daily paper in the UK. Its left-wing politics and the working-class tradition had strong connections to Moscow. It was based on a paper with the same name published in New York since 1924 under the auspices of the Communist Party USA (CPUSA). Internal disagreements over Moscow's suppression of the Hungarian uprising in 1956 caused severe disruption within the CPUSA. By 1961, the *Daily Worker* in USA had ceased, and the CPUSA itself barely survived with a fraction of its original membership.

In Britain the *Daily Worker*'s headlines, stories, editorials, even adverts, inevitably reflected its raw left wing and pro-Soviet stance. During the period of Gagarin's visit the local press, of all political colours, published details of his itinerary, sometimes including maps. The *Daily Worker* consistently published details of Gagarin's schedule not just to inform but with an obvious intent to encourage.

3. Like Some New Columbus

Arrival

Gagarin received a formal British acknowledgement of his achievement even before his plane landed. A five-member delegation of the Glasgow Trades Council was returning to Britain on the same flight after a two-week visit to the Soviet Union. The trade council secretary John Johnston presented Gagarin with a McGregor tartan tie and a Glasgow badge before they landed.[3] The Aeroflot Tupolev TU104B arrived at Heathrow just after 11:00 a.m. amidst bright sunshine, but dark clouds low on the horizon heralded the developing rain.

The question of who should welcome Gagarin on arrival had finally been settled. The influential opinion of Sir Frank Roberts from the Foreign Office that the occasion warranted a minister to welcome the spaceman at the airport was not heeded. It was the secretary of the office of the Minister for Science Francis Turnbull who welcomed Gagarin at the airport. He was accompanied by Sir Ronald Lees, the Chief of the Air Staff, and Lord Drogheda, chairman of Industrial Trade Fairs Limited. The concern that this politically weak reception party might be taken as a snub was mitigated by the announcement, just prior to Gagarin's arrival, that the Prime Minister would meet Gagarin on the Thursday, and later whilst at Earl's Court he received an invitation from the Queen for lunch at Buckingham Palace for the following Friday.

In bright sunshine, Gagarin descended the stairs of the silver Russian aircraft towards a crowd on the tarmac where he was welcomed by Francis Turnbull with the words "*Your courage and daring on undertaking the first spaceflight has been greatly admired by the people of Britain*".[4] Wearing a khaki cap and waisted jacket with Air Force blue trousers, he climbed into the open top Rolls Royce on the tarmac with the especially commissioned number plate of YG1 that would take him, with an escort of four police motor bikes, to the Soviet Embassy in Kensington.

3. Like Some New Columbus

This number plate had been commissioned with surprising haste by the organisers of the trade fair, Industrial and Trade Fairs Limited, against the advice of the Foreign Office,[5] on the grounds that even President Kennedy and his wife who had visited during the previous month did not have one. Even though the number YG1 had been sold to a celebrity for £5.00, staff from the Industrial Trade Fairs Limited had contacted the duty officer at the Ministry of Transport and acquired a response that they considered to be affirmative. The Ministry of Transport and the Foreign Office to their surprise and disappointment first saw the number plate on the tarmac at the airport, by when it was too late to intervene.

Figure 6 Nikolai Kamanin, Boris Belitzky, Yuri Gagarin and Soviet Ambassador Alexander Soldatov at Heathrow Airport, London (Courtesy Patricia Mannarino)

Patricia Mannarino, now married and based in Fort Lauderdale in Florida but originally from Wallsend on Tyne, worked in the public relations department of Industrial and Trade Fairs Limited. She had travelled throughout Britain including Manchester fulfilling public relations duties for her employer. She had also been involved in the preparation of the British Trade Fairs in Moscow's Sokoliniki Park in May 1961 where a boisterous Khrushchev attended on the opening night, commenting on everything including a win by Aston Villa

3. Like Some New Columbus

football club a few days earlier.[6] Mannarino did not attend the British fair in Moscow but her role at Earl's Court brought her in close regular contact with Gagarin and especially his interpreter, Boris Belitzky. When recalling the acquisition of the personalised number plate, YG1, for the Rolls Royce that Gagarin used during his time in London, she mused "*My boss James Brewster arranged for the YG1 number plate, we were in Public Relations and nothing was impossible*".[7]

Despite the short notice (his visit was announced only four days earlier), Gagarin drove in his open top Rolls Royce with the Soviet flag on the bonnet along streets in Hammersmith and Kensington lined with surprisingly large crowds cheering, waving and shouting with the hysteria of the yet to arrive phenomenon of Beatlemania. Young girls, foreign students, mothers with prams, Londoners and tourists overflowed from the pavements as the traffic was diverted along the route. Office and factory workers paused and took the opportunity of any vantage point, a window, balcony or a doorstep, as Gagarin drove by. This surprisingly warm welcome was repeated throughout his five days in Britain, even during the heavy rain that greeted him in Manchester.

Following a brief stop at the embassy, Gagarin arrived at the nearby Earl's Court just after 1:00 p.m., only half an hour behind schedule. The police forced a path through the dense crowd in the foyer of Earl's Court so that he could make his lunch appointment, hosted by the London Chamber of Commerce.[8] Julian Amery from the Air Ministry welcomed the Major saying "*I know that I shall be speaking for the whole nation in expressing our unbounded admiration of the coolness, skill and yet modesty with which like some new Columbus you ventured into the unknown and returned*".[9]

London Press Conference

One of the largest halls of the trade fair, housing the fashion

3. Like Some New Columbus

show, was the venue for the press conference scheduled for 3:00 p.m. With the exception of his visit to Finland earlier in July, this would be the first opportunity for western and international journalists, TV news and academics to question Gagarin about his flight directly. As with Gagarin's first press conference in Moscow on April 14th, no meaningful technical information was revealed. This was partially due to the tenacious Soviet appetite for information control but also because of the occasionally banal and trivial nature of many of the questions asked during the hour-long session. What did he think of science fiction? Did he speak English? Did he have nightmares?

A journalist from *Flight* magazine asked the one question he dreaded: *"had he remained inside the spaceship during descent or used the ejection system?"* Gagarin replied, *"he had descended in the spaceship although it would have been possible for him to eject."*[10] The BBC aerospace correspondent Reginald Turnill[11] during the Moscow press conference on 14th April 1961 had also asked him this question. His government had forced him to lie then, so he had no choice but to lie again. Here in London, his response was more ambiguous than in Moscow, indicating that he did have an option to eject. It was widely speculated at the time and formally acknowledged in 1971[12] that he had not descended to the ground in the spaceship but ejected at an altitude of 7 km.

The method of landing was an important indication of the sophistication of the Soviet technology and integral to Gagarin's subsequent claim for the world altitude record. The International Aeronautical Federation tasked with endorsing that claim required that the pilot landed in the same craft in which he had taken off.

This suspicion was further strengthened by Gherman Titov's comments during the press conference following his landing after the day-long mission in Earth orbit on 7th August 1961. Titov preserved the ambiguity by publicly declaring, in the way of a flip-side to Gagarin's response, that he *"had a choice of*

3. Like Some New Columbus

riding his capsule all the way to Earth, or of parachuting out once he dropped low enough. He chose to eject".[13]

For the secrecy-obsessed Soviets, it was unthinkable that a spacecraft launched from the Soviet Union would land on foreign soil. The Vostok spacecraft had to land in mainland Soviet Union and thus on land. The flight characteristics of the spacecraft, built in record time, would not allow sufficient loss of speed between re-entry and the Earth's surface for a guaranteed safe soft landing. The final test flight prior to Gagarin's took place on 25th March. Its passenger was a dog Zvezdockha that survived the landing without any obvious ill effects.[14] Titov was correct: it would have been possible, perhaps with some discomfort, even injury, to land inside the spacecraft. Ideally, Vostok required additional retrorockets to reduce further the speed prior to landing, but there was not enough time.

The Americans had announced their intentions to launch a person into space at the press conference on 9th April 1959, thus limiting the time the Soviets had to introduce additional modifications and the associated testing. The ejection seat seemed to be a natural solution. It offers escape in many emergencies and is familiar to all jet pilots. The spaceship that followed Vostok, Voskhod, would have additional retrorockets for a soft landing. It had to.

Voskhod was also designed and tested in haste to beat the American two-man Gemini programme. Voskhod epitomised the extraordinary risk-taking driven by fear of not being first. In an extremely speedy and equally risky modification, this three-cosmonaut spacecraft was put into space for a daylong mission in October 1964. The Voskhod was a simple and dangerous redesign of the one-person Vostok spacecraft, and without space suits and ejection seats, three tightly packed cosmonauts could just about fit in. Convinced that the Soviets had produced a fully functioning three-man spacecraft, the Americans accelerated the development of their three-man spacecraft Apollo. The Voskhod flew twice and was then abandoned.

3. Like Some New Columbus

In the period between Gagarin's flight on 12th April and his arrival in London on 11th July 1961, the American Alan Shepard had made his sub-orbital flight for NASA. That launch, like all NASA space launches, was broadcast on live television. The Soviets' secrecy was compared with NASA openness in another question from a *Daily Sketch* journalist who asked, *"whether on his next space flight, the Major would like the world to watch the attempt, for example by televising the launch?"*[15] Gagarin sidestepped the question, saying *"a colour film of the launch had been produced"*. Another questioner inquired what the medics had discovered in the post-flight medical checks. Gagarin responded that the doctors did not find any signs of physical or mental ill effects from the acceleration of launch and that since his return he slept normally. There were no residual effects from weightlessness or the strong forces during re-entry, and he concluded that humans and spaceflight were compatible. When asked about making further spaceflights he responded, *"The question of whether I shall fly to outer space again or be given some other work has not yet been decided. As far as I am concerned I would very much like to make more flights, I look forward to making them and I hope in the near future I am entrusted with additional flights"*.

A female journalist enquired whether a woman might be sent into outer space. Gagarin replied, *"I see no objections to a woman making spaceflights. A physically well-trained woman could easily stand up to the stresses and strains of such a flight. In view of the equality of men and women in Russia, there should be no objection at all in making such a flight. A woman's appreciation of beauty is even more developed than that of a man. If a flight seems beautiful to a man, it would seem even more so to a woman"*.[16] Gagarin was born in the Smolensk region of western Russia. He was a Russian and frequently slipped into using the term Russian rather than Soviet. A woman, Valentina Tereshkova, flew the last of the six Vostok flights in 1963, which was primarily driven by Nikita Khrushchev's pursuit of "propaganda firsts". Until then, all Vostok cosmonauts had been pilots; she was not but was selected for her skills as a parachutist.

An Ordinary Mortal

A writer from a culture magazine asked a question on a topic which seemed to approach the subject of spirituality. He asked if Gagarin had experienced *"what in oriental countries we call eternity?"* Somehow Belitzky's translation carried the atheist Major's bewilderment as well as the simple negative response. A representative from Associated Television proposed that the Major could never be the same again following his flight and subsequent publicity. Despite being feted by the world, mobbed by thousands during the few hours of his arrival in Britain, without a hint of irony or contradiction, and very much in line with Soviet ideology, Gagarin asserted his disagreement with this division of people between ordinary mortals and celebrities. He went on: *"We in the Soviet Union have many people who have accomplished a great deal. The number on this gold star"*, pointing to the Hero of the Soviet Union medal on his own chest, *"is numbered 11,175 which means that 11,174 people before me had accomplished something very notable"*.[17] He concluded that the discomfort of celebrity status would ease as the numbers of those who experienced spaceflight also increased.

Gagarin had joined the air force at the age of twenty-one; six years later he was fielding questions from western press with extraordinary confidence. Although surprisingly well read, he had not had an opportunity to visit a country outside the Soviet Union, but by the age of twenty-seven he had gone round the Earth as no one had before. Gagarin's mother despite her humble education had fostered in him an interest in books.

Throughout his authorized, which in this instance implies highly censored, autobiography, *Road to the Stars*, he recalls numerous books he had read during his life, including Russian authors like

3. Like Some New Columbus

Maxim Gorky, Lev Tolstoy and Alexander Pushkin, as well as authors on the international stage like William Shakespeare, Victor Hugo and Charles Dickens. The war had interrupted his education, for which he would later make up through reading as a pleasure pursuit and further study during the evenings. He was fascinated by science fiction and had read Jules Verne, Conan Doyle and H.G. Wells. He recalls[18] that H.G. Wells had visited Moscow and met Lenin. He states he was aware of Wells' book *Russia in the Shadows* and unsuccessfully attempted to get a copy from the Saratov Technical School library where he was studying. His friend and colleague Alexi Leonov who became the first to make a spacewalk on 18th March 1965 recalls meeting Gagarin for the first time in a military hospital in Moscow in October 1959, the Cosmonaut Training Centre had yet to be built. On that first meeting, Leonov recalls, Gagarin was reading a copy of *The Old Man and the Sea* by Ernest Hemingway.[19] On his own visit to Cuba in 1965, Leonov met Hemingway and personally told him that this novel had been a favourite of Gagarin's.

It is interesting to speculate what Gagarin made of the people he met in Britain and elsewhere whilst he himself was living in the goldfish bowl of the world's first spaceman. No record of his thoughts survive and it is unlikely the Soviet regime would have published a frank discourse. Those he met throughout his visit would echo the contemporary accounts of his demeanour on that day. He was handsome, with an incessant smile, engaging personality, a pleasant sense of humour and, although a confident public speaker, he was more at ease when talking to engineers, scientists and pilots rather than politicians or journalists. Outside formal speaking engagements, he was relaxed when meeting members of the public and even seemed to enjoy it.

In its 78-year history, the BIS has awarded At the end of the press conference, he received a gold medal from Dr W.R. Maxwell, the president of the British Interplanetary Society (BIS). In the brief presentation ceremony, Maxwell said, "*the*

3. Like Some New Columbus

whole world would be proud of Major Gagarin's achievement since men from all nations from the beginning of civilisation have contributed to the knowledge which ultimately made it possible".[20] Probably the world's oldest independent organisation devoted to space exploration, the BIS was founded in 1933 to promote exploration of space and astronautics through public lectures, symposia and publications.

Figure 7 Yuri Gagarin with gold medal from the British Interplanetary Society (Courtesy BIS)

One former member and chairman, Patrick Moore, host of the long-running BBC TV programme *Sky at Night*, recalls that the commission to write his first book came from a publisher from New York, following an announcement of his talk at the BIS in 1950.[21] The book was published and eventually ran to eight editions and launched not only his career as an author but also as a popular radio and television broadcaster. Arthur C. Clarke was also an early member and chairman of the BIS and went on to

3. Like Some New Columbus

write several science fiction classics of the twentieth century and is credited as the inventor of the principle of global satellite communication system using geosynchronous orbits.

the gold medal five times. In 1961, in addition to Gagarin, the BIS awarded one to Wernher von Braun. Under Hitler's direction, the German von Braun had developed the V1 and V2 bombs that had brought death and destruction to England in the closing months of the Second World War. However odd it may have looked to the outside, the BIS, an apolitical association, had already elected von Braun an honorary Fellow as early as 1949.[22]

Writing in 1961, Purdy and Burchett offer an insight into how advanced the German rocket programme was. An unnamed member of the German rocket team, then living in Spain, claims that the war had ended soon after they had perfected the V2s but their rocket programme was designed to go up to V12. V3 and V4 would have had the longer-range necessary to reach Washington and Moscow. He speculated that from V6 onwards the rocket programme was a space programme, insisting that the Redstone Rocket that launched the first American into space was in effect a V6 rocket.[23]

Subsequently von Braun masterminded the programme that ultimately fulfilled President Kennedy's goal of *"landing a man on the Moon and returning him safely to the Earth"*.[24] The decision for a British organisation to award von Braun a medal inevitably drew criticism. The *Daily Worker* editorial under the heading "Double Cross?" insisted that this idea *"should be dropped at once"*.[25] It was not, and von Braun got his medal.

In 1964, the BIS awarded its third gold medal to the first female in space, Valentina Tereshkova. The attempt to award gold medals to the Apollo 11 astronauts was not initially successful. Although the medals, complete with the inscriptions were ready prior to the actual Moon landing, during their fleeting visit through London, Neil Armstrong, Buzz Aldrin and Michael Collins did not have time to be personally presented with the

3. Like Some New Columbus

medals. With extreme disappointment, the BIS had to wait for an alternative opportunity. The astronauts were not scheduled to visit the UK. They could post the medals but that was unthinkable. On 9th of February 1970, NASA administrator Thomas Paine came to London and accepted the gold medals at a special meeting convened by the BIS. On his return to the USA, he presented the medals to the three Apollo Astronauts on 17th February 1970.[26]

Figure 8 James Brewster with Yuri Gagarin at the Soviet Embassy
(Courtesy Patricia Mannarino)

At the end of the press conference, the fashion hall and surrounding stairs and corridors were jammed with journalists and spectators. The density of crowd would make Gagarin's attempt to tour the Trade Fair impossible. At one point as he came through "cosmic hall" followed by a crowd of photographers and journalists, congestion and disorder ensued. The standalone display of a life-size Sputnik was knocked to the floor without damage or injuries, but it was clear that his presence had brought the Trade Fair to a standstill.

3. Like Some New Columbus

Consequently, Gagarin left Earl's Court so abruptly that his car was not available outside the entrance.

The final item on Gagarin's itinerary for his first day was another formal reception hosted at the Soviet Embassy itself. Here were so many guests that they were formally welcomed in two separate shifts, at 5:30 p.m. and again at 7:30 p.m.[27] Gagarin along with the Soviet ambassador received hundreds of guests from all occupations and political allegiances.[28] They included Reginald Maudling, the president of the Board of Trade, David Eccles, the Minister for Education, Ted Hill, the chairman of the Trade Union Council, and John Gollan, the leader of the Communist Party of Great Britain. Also present were Hugh Gaitskell, the leader of the Labour Party, and Harold Wilson who succeeded him and following a general election victory three years later became Prime Minister.

4. Together Moulding a Better World

Before becoming a global household name as the world's first cosmonaut Gagarin had been immersed for four years in the role of a foundryman. Eventually qualified as a caster, he had first-hand experience of the dirt, grime and danger of working with high-temperature furnaces and handling molten metal in a foundry. That personal experience of the factory floor combined with the first-hand experience in his own family had nurtured in him a deep sense of empathy for working class values which brought him to Manchester.

The twelfth of July 1961 was one of those hot summer days where sharp, heavy showers alternated with bright hot sunshine. It was surprising there was no thunder and lightning to greet the cosmonaut as he arrived in Manchester, there so easily could have been. The front-page headlines in the *Manchester Evening News* on the following day (13th July) read "Monsoon and Gales hit Northwest". Wigan, Leigh and Birkdale experienced local flooding, and graduates at the University of Manchester had to hold on to their gowns and mortar boards during the annual graduate awards ceremony at the Oxford Road campus.

Although dates and times for Gagarin's meeting with the Prime Minister and the Queen had been established with some degree of certainty, other meetings tended to be last minute arrangements. Although the invitation to Manchester from the Amalgamated Union of Foundry Workers (AUFW) had been formally accepted on 23rd May 1961, the arrangements to visit Manchester on the 12th July was only finally confirmed on the 11th July by a telegram, marked "priority", from the Soviet Ambassador Alexander Soldatov to the Lord Mayor of Manchester Lionel Briggs.[1] Gagarin was scheduled to arrive in Manchester on the 12th at about 10:00 a.m. at Manchester's Ringway Airport (now Manchester International Airport) and visit his hosts' Union head quarters in Old Trafford, an the

4. Together Moulding a Better World

Metropolitan Vickers engineering works in Trafford Park and then the Town Hall, before returning to the airport six hours later for his return flight to London.

The immediate aftermath of Gagarin's successful flight made him a household name overnight. Details of his flight, the spacecraft and his personal life were on the front pages of newspapers around the world. When the Executive Council of the AUFW discovered that he had been a moulder, they decided it was appropriate to invite him to Manchester and offer him an honorary membership of their union in recognition of his "*contribution to the progress of humanity*".[2] During the Annual Delegate Meeting on 23rd May 1961 at the seaside resort of Great Yarmouth, the Executive Council reported that Major Gagarin had accepted the invitation to visit Manchester and an honorary membership of their union.[3]

The AUFW played a central role in preparing Gagarin's visit to Manchester. Fred Hollingsworth, president of the AUFW, went to London to greet Gagarin on his arrival in Britain on the 11th July and flew with him from London to Manchester on the 12th. Also on-board were ambassador Soldatov, Mr Francis Turnbull, Gagarin's overseer Nikolai Kamanin, his translator Boris Belitzky, and another party of fifteen mostly Soviet journalists.

A specially chartered Discovery Class twin-engine turboprop BEA Viscount 800 left London Heathrow where the rain had already started. At a time when it was common for train engines to have names, as ships still do, this three-year-old aircraft also had a name. The aircraft registered as G-ADYN had the fitting name "The Sir Isaac Newton".[4]

The pilot Captain Stanley Key invited Gagarin to take the co-pilot Brian Long's seat for six minutes during the flight to Manchester. The steward Michael Swiety made the announcements in fluent Russian during the flight, while refreshments of sandwiches and caviar were served by

4. Together Moulding a Better World

stewardess Monica Fornara in the first-class section at the back of the aircraft. With a broad smile, Gagarin chose the caviar.[5]

Figure 9 Yuri Gagarin in co-pilot's seat during flight to Manchester
(Courtesy Marx Memorial Library)

The whole of the forty-minute flight at 10,000ft was however marred by poor visibility, and although Gagarin had a window seat, he could not make out much detail during his only opportunity to see rural and industrial England from the air. By the end of the day, exhausted by the hectic schedule, he would quickly fall asleep on his return flight to London.[6]

It was raining when he arrived at Manchester's Ringway Airport, but only a few in the crowd waiting for him took the trouble to use umbrellas. The crowd at the airport in Manchester was larger than the one that had seen him off about an hour earlier at Heathrow. The aircraft taxied close to a hangar that had been hurriedly prepared for an impromptu welcome ceremony. Before descending, Gagarin waved and saluted from the top of the aircraft steps. A spontaneous applause erupted as he descended wearing a military style cape over his khaki uniform, with the Soviet ambassador immediately behind him. In traditional Soviet style, Gagarin applauded back.

4. Together Moulding a Better World

Despite the rain, the enthusiastic crowd was becoming too big for the number of police tasked with keeping order. A short formal welcome by the Lord Mayor of Manchester, Alderman Lionel Biggs in top hat, tails and chains of office, was conducted in a hastily organised reception area within the large Hangar number three.

Figure 10 Hangar #3 at Manchester Ringway Airport
(Courtesy Fred Garner)

With the grey-haired Fred Hollingsworth sitting on one side and the Soviet ambassador on the other, Gagarin stood in the middle as the open top beige Bentley attempted to drive through the good-natured but out of control crowd. Linking hands, the police cleared a path for Gagarin's slow drive out of the airport. The Bentley had been provided by Rolls Royce through their local agent Cockshoot's and was driven out of the hangar by Ernest Morris, a fifty-year-old former locomotive driver for British Railways from North Manchester.[7] Cockshoot's, on the understanding that Gagarin would be accompanied by his wife, had sent a second car, a dark blue Rolls Royce Silver Cloud. She was not with him on this trip but the car was used by other members of the party. He drove out of the airport standing in his car, but as the crowd density fell, he sat down, and the roof was closed.

4. Together Moulding a Better World

Gagarin's route, unchanged despite the weather, would take him along Ringway Road, Shadow Moss Road, Brinley Road, Altrincham Road, Princess Parkway, Barlow Moor Road, Upper Chorlton Road and then a short distance into Chorlton Road.[8] After a couple of kilometres, although fewer in number, people were still standing along the roadside cheering and waving as his convoy passed. Eventually, Gagarin asked for the roof to be removed once more. He stood up and waved back all the way to Old Trafford. His military cap and cape gave him some protection against the rain, but the rest of the elegantly dressed passengers in his car got a soaking.

A little bit of History

Of the thousands amongst the crowds who lined the route in the rain that day, there were two mothers with their babies. Patricia Hayes along with her friend pushed their prams from Heald Green to Shadow Moss Road to see this "*little bit of history*".[9] It was still raining and on that part of the route, Patricia remembers, there were only a handful of sightseers. The most striking thing about Gagarin she recalls was that "*he looked incredibly young*".

Yew Tree Comprehensive School was a progressive and only the second comprehensive school in the Manchester area in 1961. The headmaster, Mr. Davis, during the morning assembly on Wednesday 12th July, had forbidden his students to go out to see the cosmonaut driving along Princess Parkway, the main road connecting central Manchester and the Airport. He offered no reason for this decision at the time, but his motives became clear many years later.

At fifteen years of age, Keith Brown, a Yew Tree School student, had often been, as he recalls, "*frog marched*" out to Princess Parkway to wave at passing royalty and VIPs and so was very surprised by the headmaster's decision during the morning

4. Together Moulding a Better World

assembly. Years later on a chance meeting with his former headmaster, Keith Brown asked the question which he could not have dared to put at the time. Why were they not free to greet the cosmonaut? Long since retired, the headmaster was also free to offer the answer that he would not have been able to share at the time. He explained, "*it was the then Conservative City Council's directive and not his*".[10]

Between the school and the Parkway, however, was the school's playing field. A well-trodden route led to a break in a line of bushes that allowed anyone to get to the Parkway from the playing fields. Wendy Jones (nee Hamilton) was in a Mathematics lesson that morning, but her teacher Mr. Greenlees did not want to miss out. He escorted Wendy and the rest of the class across the playing field and through that hole in the bushes onto the Parkway to see the cosmonaut. Gagarin, still in his car with the top open, drove by along with his convoy. She recalls him "*waving to everyone. Our teacher said it was a moment we would always remember, and he was right*".[11]

Doreen Williams (nee Aldred) was also at Yew Tree at the time and recalls that prefects were posted on the main entrance to prevent students from sneaking out. But she like so many found her way through the same break in the bushes from the playing field leading out to the Princess Parkway to see the spaceman. She remembers watching Gagarin drive by, along with "*hundreds of others*". Doreen does not remember the exact punishment for this misdemeanour on their return, but given the large numbers that had disobeyed the headmaster's instructions, it was not severe, "*probably only a talking to*", she concluded.

In contrast to Yew Tree School's headmaster, the Manchester Lord Mayor honoured the Soviet Major with a civic reception at the Town Hall in his honour as his final stop before returning to the airport for his flight back to London. Many dignitaries of varying political colours would also be present at the Town Hall for the afternoon reception. One who would meet him again at

4. Together Moulding a Better World

the Town Hall was waiting amongst the crowd on the junction of Princess Parkway and Barlow Moor Road.

Dame Kathleen Ollerenshaw, a mathematician by profession with a long history of public service, was ideally placed to meet Gagarin twice. During 1975-6, she served as Manchester's Lord Mayor, but at the time of Gagarin's visit, she was a Conservative local councillor for the district of Rusholme. At the time, she lived close to Barlow Moor Road, where at the age of ninety-eight she continues to reside. An experienced amateur astronomer, she was interested in Gagarin's visit and keen to see his progress through Manchester near her home, and ensure her children participated in the experience. This interest in astronomy and space exploration never left her: she followed the work of Sir Bernard Lovell in building and operating the Jodrell Bank 76m diameter Radio Telescope thirty kilometres south of Manchester. During her busy tenure as the Lord Mayor of Manchester, she fondly recalls the unplanned precarious but exhilarating visit to the inside of the telescope dish while it was undergoing maintenance during a visit in 1976 on a cold February morning.[12]

On the morning of Gagarin's visit, she had only a short walk from her home to join the crowds waiting at the junction of Barlow Moor Road and Princess Parkway. Both her children, nine-year-old Charles and five-year-old Lawrence, accompanied her, but neither shared her enthusiasm for this encounter. She had waited for about an hour before Gagarin drove by. She recalls with passion the significance of the event as she reminded both her children "*there is somebody, the first human being ever to get away from the Earth.*"[13] Many people in Manchester must have felt the same sense of awe and elation at this one chance of seeing Gagarin. Once Gagarin's convoy and the police outriders had swiftly turned left from Princess Parkway onto Barlow Moor Road, she rushed home, dropped off her children and headed off in her car to the Town Hall for her second meeting.

4. Together Moulding a Better World

As a local councillor, she had already arranged to be amongst the dignitaries who welcomed Gagarin for the official civic reception in the gothic architecture of the Town Hall in the centre of Manchester following his visit to Trafford Park. Later, on arrival in the Town Hall, Gagarin shook the hands of several guests, including hers as he entered. She recalls with great thrill *"When he got to me I remember looking at this clean-shaven young man and saying "wonderful". It is the only word I remember hearing myself saying, it is not something you normally say. He was very short, smart in his khaki coloured uniform and without his hat at this time."*[14]

As his motorcade turned left into Barlow Moor Road, he was about 5 km from the AUFW HQ in Old Trafford. For the next couple of hundred metres on the right-hand side, the convoy passed Manchester's large municipal cemetery with its rich working class and aviation history. The cemetery serves the variety of communities that have over time settled in Manchester, and it is the final resting place for many who played a key role in bringing the city to a wider international recognition. Sir Matt Busby, the revered manager of Manchester United Football team, and the artist L.S. Lowry who captured Manchester's working class and industrial heritage through his painting of "matchstick" figures would find their last resting place there several years after Gagarin's visit. Sir John Alcock (born in Stretford Memorial Hospital a few kilometres along the road Gagarin was travelling) is buried in Southern Cemetery. It is possible that Gagarin would have come across Alcock during his early days as an aviator. Like Gagarin, Alcock was also an adventurer and chose to "push the envelope" in aviation.

Manchester and Aviation

Opened in 1938, Manchester International Airport is the largest regional airport serving over 20 million passengers every year. It has two parallel runways, each three kilometres long, employs

4. Together Moulding a Better World

almost twenty thousand people on site, and indirectly supports the employment of another forty thousand. There are another two active airports in use with the Manchester area today – City Airport Manchester (formerly and still popularly known as Barton Aerodrome) and Woodford Aerodrome in South Manchester, but in the past, there were an additional three in Trafford Park, Chorlton and Wythenshawe, no signs of which now remain.

The first purpose built an airport in Manchester was the Trafford Park Aerodrome opened in 1911, located in Trafford Park close to where a large shopping centre (the Trafford Centre) now stands. Today Tenax Road runs across the site of where the large grass runway was located. It operated through the First World War and closed in 1918. It serviced biplanes with small engines, the earliest aeroplanes that flew over Manchester. It was nothing more than an open field, but it inspired the early aviators in Manchester including Alliot Verdon Roe and John Alcock who both flew there.

Alexandra Park Aerodrome opened in the same year that Trafford Park Aerodrome closed. It also operated a grass landing strip. Gliders, towed by cars, used the airfield alongside powered aircraft. A.V. Roe established his company, the Avro Transport Company, at Woodford Aerodrome following the closure in 1924 of the Alexandra Aerodrome. The Lancashire Aero Club (LAC), now the oldest established aero club in the UK, was founded at the Alexandra Park Aerodrome in 1924 and moved with Avro Transport Company to Woodford Aerodrome where it was based until the outbreak of the Second World War when it moved to Barton Aerodrome.

Barton Aerodrome, completed in 1930, was for a few years a serious contender for the main city airport. In the end, Barton Aerodrome was considered too small, and Ringway was selected as the site for the main airport where the Manchester International Airport stands today.

4. Together Moulding a Better World

While Barton Aerodrome was under construction, a temporary airfield, used between April and December 1929, was constructed on the site of four fields of Rackhouse Farm. Those fields like that of Trafford Park and Alexandra Aerodrome are now roads, businesses and housing. Lancashire Aero Club continues with its rich tradition of aviation in Manchester, operating from Barton Aerodrome until 2009 when it moved to a single grass strip airfield at Kenyon Hall Farm, near Wigan. Today, BAE Systems operate Woodford Aerodrome, and with the recent cancellation of the Nimrod aircraft the aerodrome now has an uncertain future.

Along with Lieutenant Arthur Whitten Brown, Captain John Alcock had made the first non-stop flight across the Atlantic. Using a modified Vimy IV twin-engined biplane, they took off from Newfoundland on an overcast afternoon of 14th June 1919, and after a journey of almost 2000 nautical miles over sixteen and half hours they landed in Ireland on the following day. A testimony to their skill, courage and tenacity was the success with which they negotiated a series of perilous challenges, any one of which could have led to a sudden and catastrophic end to their adventure and lives. Dense low-level fog, high-level cloud, engine problems and frequent failure of instruments constantly threatened the flight. The severest of challenges came from an unexpected and unpredicted snowstorm causing snow blockage on the engine intakes. Without their engines, they would have to ditch. The only solution was to leave the cockpit and walk along the wing to the engine to manually unblock the intake, walk back and repeat the entire risky process on the other wing and engine. Astonishingly, as the snow persisted, with Alcock at the controls in the midst of the raging storm, Brown had to repeat the dangerous wing-walking feat four times, in the dark. Dawn broke at 06:20 a.m. and they landed safely thirty minutes later in Clifden in Ireland.

Alcock died before the year was out whilst delivering a flying boat from England to the Paris Air Show on 18th December

4. Together Moulding a Better World

1919. He had been born in Stretford Memorial Hospital, about five minutes' drive down the same road from where he is buried. His unique contribution during his short 27 years had taken aviation innovation and development further than this short 3 km journey from his cradle to the grave. It is fitting that Gagarin, a 27-year-old champion aviator, passed close by the grave of another. Seven years later, like Alcock, Gagarin, too, would die in the pursuit of his passion of aviation.

Manchester is well known for the world's first intercity passenger railway running between Liverpool and Manchester, opened on 15th September 1830. But not many are aware that Britain's first scheduled domestic air service had once been based at Alexandra Park Aerodrome located on the other side of the cemetery he was passing. Operated by the Avro Transport Company, a three-seater Avro 504 biplane made the forty-five-minute flight from Manchester to Blackpool via Southport between 24th May and 30th September 1919. The Alexandra Park Aerodrome operated between 1917 and 1924; a housing estate and Princess Parkway built through the middle of the former landing strip in the following year dramatically changed the landscape, permanently obscuring its former role. On the site where once the aerodrome hangars and associated buildings were erected, today stands the Greater Manchester Police Sports and Social Club.

As Gagarin's convoy headed south, nine-year-old Pauline McLaughlin lined up outside Chorlton Park School with her classmates on Barlow Moor Road. Almost fifty years on she has not forgotten, *"The day was very exciting, we were going to see someone who had been up in space. It remains a clear memory, and I have never forgotten his name, useful in a pub quiz"*.[15] The centre of Chorlton today is very different to that of 1961. Today it has an eclectic mix of a bohemian culture popular amongst university students who have chosen to make home there, with outdoor seating for cafés, restaurants and pubs. Barlow Moor Road briefly changes its name to Manchester Road and then

4. Together Moulding a Better World

again, to become Seymour Grove. Prior to Seymour Grove, Gagarin's motorcade turned right into Upper Chorlton Road, where the children of Manley Park Junior School were waiting along with their teachers.

Exactly a hundred years before Gagarin's arrival, Sir James Worrall had been the mayor of nearby Salford and later built his family house on Upper Chorlton Road. He called it Chrimsworth House after the dyeing company he owned in Hebden Bridge, Yorkshire. It was a grand house complete with stables that survived until 1969, the year that Buzz Aldrin and Neil Armstrong took humanity's first step on the surface of the Moon. It had been converted for use as a kindergarten teacher training school, and in July 1961 it was known as the Chrimsworth Annexe to Manley Park Junior School. The stables had long been converted into play areas for children, and it was outside these former stable gates which opened out to Upper Chorlton Road at about 10:30 a.m. that Gagarin's convoy would pass.

A seventeen-year-old Marjorie Rose was attending Manley Park every alternative week as a trainee teacher for kindergarten. The school children and staff, around eighty in total, joined the crowds already lining the road. They were waving red flannel dusters they had made into red flags as he drove by. Marjorie recalls the weather to be damp and miserable but there was no heavy rain as Gagarin drove by, to cheers and energetic flag waving. The general traffic had been stopped by the police motorbike riders so Gagarin's car drove by unhindered. Almost fifty years on, Marjorie avidly recalls, *"he was in an open top car and had his, what seemed rather large, hat on, but those gorgeous eyes, ooh I go all goosepimply even after all these years, and he was as excited as we were. We all waved and cheered back at full throttle"*.[16] Marjorie's mother had been working at a hotel close to the airport and reported a similar exuberant welcome for the only person who had been in Earth orbit.

4. Together Moulding a Better World

This was not Marjorie's first encounter with Soviet history in Manchester. Despite the Cold War enmities between governments, there were many instances where former victims of Soviet persecution put aside any bitterness or ill feeling and joined into celebrate this unique human achievement despite it being primarily a Soviet feat. As part of their training, all trainee teachers were required to conduct a research project focusing on one of their students. Marjorie chose a four-year old Alicia Kosenko. Alicia's parents were two of the 200,000 Hungarian refugees who had fled Hungary following the failed uprising in the autumn of 1956. After an attempt to oust the Soviet Union imposed government, an overwhelming Soviet military force crushed a spontaneous revolt by ordinary Hungarians. Ultimately, the revolt failed with losses on both sides. Around 2,500 Hungarians and almost a thousand Soviet troops were killed during the peak of the uprising between 23rd October and 10th November 1956. Despite their suffering, genuine good will and sincere admiration of Gagarin's achievement would be exhibited by other former victims of the Soviet persecution in the crowds of Manchester that day.

In the absence of concrete evidence, Marjorie is uncertain, but she may have been a victim of her own communist connections. After completing her training, Marjorie applied for a post as a nanny in the USA. Despite her qualifications and a glowing reference from her vicar, she was unsuccessful. She would eventually discover through her vicar that her affiliations with the British Communist Party were probably the grounds behind her rejection. She was not a member of the British Communist Party, but her late father had been.

Gagarin's convoy was now moments from the head office of the Amalgamated Union of Foundry Workers, where, despite the rain, large crowds spilling onto the main road had brought the traffic to a halt. The cars turned left at the end of Upper Chorlton Road onto Chorlton Road and immediately stopped. In the union office, specially invited senior union officials waited for an

4. Together Moulding a Better World

improbable but magnificent, experience in the history of British Trade Unionism.

Within a few days of his historic space flight into space on 12th April 1961, Gagarin burst out of the closed Soviet society onto the world stage. Unknown to the union, Gagarin had probably the strongest affinity with the working-class traditions than any of the group of twenty pilots from which he had been selected to make the first flight. Manchester is one of the oldest industrial cities with firm roots in the industrial revolution, a large working-class population and a long history of promoting trade unionism. Gagarin shared many of those traditions and that was perhaps the motivation for his personal choice in coming to Manchester.

A day after his flight, on the 13th April, Gagarin had given one of his first interviews to an *Izvestia* correspondent. Near the end of the interview, answering a question about the future of space travel, Gagarin predicted, *"I am sure the time will come when trips around the Earth will be organized by trade unions."*[17] Born and brought up under the creed of collectivism Gagarin shared genuine and natural accord with the principles of trade unionism. His father had worked as a carpenter and his mother as a milkmaid on a local collective farm.[18] He like the rest of his extended family had been surrounded by the communist principle of joint working and working-class traditions. His personal life story had all the characteristics of a flawless, exquisite product of the Soviet system. These strong working-class credentials contributed to his selection as the first cosmonaut by Khrushchev, who saw in Gagarin a perfect propaganda tool to deploy against western capitalist societies. Khrushchev understood perhaps better than Gagarin himself that Gagarin's mission would not end once he had safely returned to Earth.

In 1949, Gagarin had left home in Smolensk region for the first time and, whilst nominally in the care of his uncle Savely

4. Together Moulding a Better World

Ivanovich in Moscow, about 200 km away, he went to study in a vocational school in Lyubertsy, just outside Moscow. He wanted to be a fitter, but being a year behind the required education, he did not qualify and accepted a place on the foundry course instead.[19]

Foundry work is not well paid; it is hard, dirty and dangerous. Casting is the process by which molten metal is poured into a mould, and, once cooled, the object (church bells, components for machinery or a ship's propeller) is removed from the mould or the mould destructively removed from it. A moulder decides how the mould is to be made, and using the appropriate material, he hollows out a cavity for the mould. Depending on the nature of the object, a core maker designs and makes the cores that introduce a cavity into the mould and thus the final product. The pouring of molten metal into the moulds containing the cores would have been predominantly manual and highly hazardous work in Gagarin's time. Modern day moulders and foundrymen perform largely the same activities, albeit in a safer, cleaner environment with a higher degree of automation.

In his highly censored autobiography (more a sycophantic celebration of Communism than an accurate record of his life), he describes his personal focus on foundry work during his time as a student. For his final year diploma project, he states, "*I had to plan a foundry for the serial production of big castings with a capacity of 9,000 tons of castings a year. In addition to that I had to plan the technology of the casting and methods of teaching their production at the technical school.*"[20]

In the spring of 1955, his fourth and final year at the Saratov Technical College, Gagarin was working on both his college diploma and his civilian flight training. By the end of the summer, he had successfully completed both.

At this point, he could have continued his career in industry or in flying. Under the guidance of his flying instructor Dmitry

Martyanov, he enrolled as a cadet with the Soviet air force in Orenburg,[21] ending his connection with foundry work forever. Six years later, his limited foundry work experience brought him to some of the largest and oldest foundries, in the birthplace of the industrial revolution, Manchester.

Working Class Traditions

The greatest social change, perhaps since the introduction of farming, were the changes introduced by the industrial revolution. Large cities with huge populations became the centres of industrialised urban societies. It was the result of several factors coming together including the development of banks, an established liberal capitalist economy, a land owning individualistic society, and cheap raw materials from the outposts of the empire. Stable health and education systems brought profound socioeconomic, cultural and eventually political changes that would instil a habit of technological development that eventually would take mankind into space. Many of these developments would in time spread around the world, but they started in England and especially in Trafford Park in Manchester.

Large-scale manufacturing demanded a large-scale labour force. The populations of Manchester and Liverpool grew through migration from Lancashire, Cheshire and Ireland. Employees in textile mills, mines, foundries and factories experienced sustained and substantial income, more than they could have earned by working the land. The land and factory owners benefitted from huge growth in their own private wealth. There were shortcomings, too, chronicled memorably, for example, by Manchester Victorian novelist Elizabeth Gaskell, but Manchester and its industries thrived, especially that of metalworking.

The job of a foundryman, and they have always tended to be

4. Together Moulding a Better World

men, has been in existence since the Iron Age. The techniques of metalworking arose in different parts of the world at different times. As industrial cities experienced large growths in their populations, they were transformed into large urban metropolitan communities. Peasants and labourers from these communities gave rise to the industrial working classes, including foundry workers.[22] It was amongst the working class people in these communities where the struggle for safer working conditions, fairer wages and a desire for equitable relationships between employer and employee arose. Subsequently this was also where the demand for the moderating influence of a trade union first emerged.

The story of trade unionism in foundry work is emblematic of the development of unionism during the Industrial Revolution. Iron smelting and manufacturing of cheaper, more efficient iron tools than bronze was the prerequisite for the infrastructure (clearing forests, draining, transport and cultivation of land on a large scale) for an industrial society. This change in the scale of production introduced a fundamental shift in how iron founders and moulders worked. They used to be self-employed, working in established small-scale operations to serve their local communities. The eighteenth century saw the introduction of the new technology of the Cupulo Furnace, which could re-melt pig iron to improve the composition and characteristics of the melt for industrial production. New fuels like coke instead of charcoal and the new blast furnace enhanced the efficiency of smelting. Demand was high, ranging from machines for factories, steam engines and ships to cooking stoves for the growing cities.

This demand could not be met by small-scale self-employed craftsmen. Large-scale production required a large-scale enterprise with investment in furnaces, buildings and many individuals to sustain the infrastructure of the enterprise itself. This considerable investment by wealthy employers introduced a fundamental change in working relationships. Moulders would become wage-earning employees working for the capitalist

4. Together Moulding a Better World

employer. In this new relationship moulders sold their labour to make products that would be sold by the employer to make profit for his investment. It was the pursuit of a balance between fair pay and reasonable working conditions for the employees and profit for the employer that gave rise to the trade union movement.

To enhance profits, employers, individually or through associations with other employers, maintained low wages for their employees and sought to maximise the price for the products. Prior to trade unions proper, Trade Clubs and Friendly Societies had been in existence not only for foundrymen but also in other industries like nail makers and stonemasons. Employees recognised that their wages were their livelihood and the only way they could protect it was to combine with other employees and form workers' unions.

Tensions developed between employees and employers when the economic cycles' highs, lows, occasionally depression, and recession occurred. It was in the shadow of such a depression and the backdrop of the Napoleonic Wars that a Foundrymen's Union was first established on 6th February 1809 in the Hand and Banner Hotel[23] in Bolton, then known as Bolton-Le-Moors, about 20 km north-west of Manchester. A group of iron moulders agreed to set up a Friendly Iron Moulders' Society (FIMS). After 1854, the union changed its name to the Friendly Society of Iron Founders (FISF). This change in name was not significant, and potentially motivated by the desire to register under the Friendly Societies Act of 1854.

During the summer of 1851, Hyde Park in London hosted the Great Exhibition. This first exhibition gave rise to a series of World's Fair Exhibitions that became popular around the world. A century and a decade later, in the faded glory of the Great Exhibition, Earl's Court hosted the Soviet Trade Fair that brought Gagarin to London. The Great Exhibition of 1851 proved to be a perfect opportunity to showcase, amongst other

4. Together Moulding a Better World

items, the work of foundrymen. Exhibits came not only from Britain but from all parts of the then expanding colonies. They included kitchen appliances, a reaping machine, the Jacquard loom, the barometer, locks, photographs, a basic fax machine and a single cast iron frame piano. Perhaps the most expensive exhibit was the then largest known diamond, the Koh-I-Noor, which Gagarin would see during his visit to the Tower of London. All of these exhibits were housed in a unique iron glass building, the Crystal Palace, an exhibit in its own right, made by moulders. At a time when Britain was producing more iron than all the other nations put together, a typical foundryman worked sixty hours a week for thirty-four shillings in Manchester. Rates varied between foundries and regions. The concept of overtime did not exist, workers did not have the vote and the employers harnessed the power of unjust law to restrict the rights of employees to negotiate their labour openly.

Over the following twenty-five years, the quality of life for foundry workers improved. By 1873, the unions had secured a shorter working week of fifty-four hours, pay rates had increased to thirty-eight shillings, overtime rates were established, and the reviled Masters and Servants Act of 1823[24] was repealed. Union membership increased from 4,512 in 1847 to 12,336 in 1875.[25]

The union that hosted Gagarin, the AUFW, came into existence in the immediate aftermath of the Second World War through the amalgamation in 1946 of three related unions – the National Union of Foundry Workers, the Iron Founding Workers' Association and the United Metal Founders Society. It grew strong quickly. In just three years, it attracted a membership of over eight thousand. In 1968, the year of Gagarin's death, the AUFW was itself amalgamated and became the Foundry Section of the Amalgamated Union of Engineering & Foundry Workers.

The AUFW headquarters was located at 164 Chorlton Road, a hundred metres from a five-road junction known colloquially as "Brooks Bar". It is named after a former tollgate guarding an

4. Together Moulding a Better World

exclusive area owned by Samuel Brooks, who in 1834 had bought 63 acres of what was then known as Jackson's Moss.[26] He had drained the moss and built exclusive villas for wealthy businessmen and a large house for himself. In deference to the town of Whalley near Blackburn where he had grown up, he named his house Whalley House. The tollgate, located on the junction of Withington Road and Upper Chorlton Road, where today stands the Whalley Hotel, stopped operating as a tollgate in the late nineteenth century.

Figure 11 Yuri Gagarin arriving at AUFW Headquarters
(Courtesy Alf Lloyd)

In the year approaching the 50th anniversary of Gagarin's visit, 164 Chorlton Road is a combined local GP surgery and a chemist. On 12th July 1961, it was the freshly painted headquarters of the Amalgamated Union of Foundry Workers, and in the boardroom on the first floor the world's first spaceman formally became the AUFW's first honorary member. He was presented with a medal, a picture of two right hands gently cupping a fragile globe, inscribed with the words "Together Moulding a Better World". The other side of the medal contained the words "*Amalgamated Union of Foundry Workers of Great Britain and Ireland*" along the outer edge, and "*Presented to Major Yuri Gagarin First Honorary Member*

4. Together Moulding a Better World

A.U.F.W. 23rd May 1961" in the centre. The date 23rd May refers to the date when the invitation was formally accepted. A six-foot diameter copy of the medal, picturing the two hands cupping the Earth, was hung above the building entrance for the occasion, together with the union and Soviet flags and welcome messages in English and Russian.[27]

Figure 12 Fred Hollingsworth presenting the AUFW gold medal
(Courtesy Alf Lloyd)

The union's decision to use the relatively small office rather than a larger venue had been by design. *"From the outset we were determined that the presentation ceremony should take place in our own trade union office amongst foundry workers. This meant disappointing many friends in the wider Trade Union Movement who wanted to attend, but it was in our view most important that Gagarin should meet our own members."*[28] Of all the public meetings, this would be the smallest of venues during Gagarin's visit to Britain.

The general secretary of the AUFW, Fred Hollingsworth, had flown out to London so that he could be with Gagarin on the flight to Manchester, and he was sharing the open top car ride in the rain from Manchester Airport to the union office. The gold medal had been designed by the publicist Ken Sprague and made by a Birmingham company, which did not participate in trade unionism.[29] Sprague was a member of the Communist Party of Great Britain (CPGB) until 1988 (it was formally disbanded in 1992), and for a time the publicity manager for the CPGB affiliated newspaper, *The Daily Worker*. Sprague, more than anyone, had promoted the importance of publicity as a tool to enhance trade unionism.

As Gagarin and his cavalcade arrived outside the union office, Adam Faith's film debut "Beat Girl", about the youth rebellion of the late fifties, was playing at the Imperial Cinema next door.[30] Crowds in raincoats and umbrellas, but some oblivious to the rain, greeted the cosmonaut as he arrived in a particularly heavy shower. In addition to the seventeen in Gagarin's party, the union HQ was going to accommodate numerous chairmen and secretaries of district committees, its general secretary David Lambert, the chairman of the TUC Ted Hill, the secretary of Manchester and Salford Trades Council Horace Newbold and Ellis Smith MP. A cleaner approached the police at the entrance and successfully asserted her right to enter because "*she worked there*".[31]

The office had been prepared to accommodate the largest number of guests it would ever have. It had been redecorated internally and much of the furniture removed to the back to accommodate the visitors. With seats for only fifty-six, over one hundred people squeezed into the first-floor boardroom to watch the presentation ceremony. Following refreshments of bread rolls, prawn, cheese, anchovies along with Russian tea and vodka, Gagarin made a short appearance at the first floor window and waved at the rain-soaked crowd outside.

Brother Gagarin

Overwhelmed with pride, Fred Hollingsworth started the ceremony to enrol the union's first honorary member in front of

4. Together Moulding a Better World

union officials, press and cameramen. He described the occasion *"as a great and historic one in British Trade Union history"* and continued, *"he felt greatly privileged in being able to pay this honour to Major Gagarin"*.[32] Gagarin was already wearing his Hero of the Soviet Union medal when Fred Hollingsworth pinned the Union's gold medal on his chest and then presented a duplicated address illuminated in red, green and yellow with English on one side and Russian on the other. A five-minute cine film[33] recovered in 1987 has two short audio sections that are the only direct record, once translated, of what Gagarin said during his time in Manchester. Fred Hollingsworth presented the medal with the words *"with a sincere hope and through a peaceful coexistence and friendship [between] our two great nations that this will be carried out. I have great pleasure on behalf of the union to present this to Brother Gagarin."*[34] Addressing him as *"Brother Yuri Gagarin"* Fred Hollingsworth continued, *"now you are a member of the Great British Trade Unions Movement and number one honorary member of the Amalgamated Union of Foundry Workers"*.

Speaking in Russian that his translator Boris Belitzky translated simultaneously, Gagarin said, *"I consider it the greatest honour possible to be an honorary member of your union, a union which ranks amongst the oldest in the world and has such fine traditions. I shall consider it the greatest privilege to wear this honorary token which is a symbol of the ideal of a better world, a world of peace and friendship"*, clapping Soviet style in response to the audience's loud applause and cheers. Throughout his five-minute speech, he stared mostly at the floor in front of him rather than the tightly packed audience in front. He went on to explain the peaceful nature of his spaceflight. *"It had purely peaceful scientific objectives. As you know, the spaceship was not carrying any weapons. It did not even carry any cameras to take pictures of any part of the world"*.[35] He had emphasised the peaceful nature of his spaceflight perhaps because he was aware more than his audience of the escalating political tensions over Cuba and especially East Berlin. The reference to the absence of

4. Together Moulding a Better World

cameras aboard his spacecraft was a thinly disguised rebuke of the failed American attempt during 1960 by the CIA controlled U2 spy plane to take pictures deep inside the USSR territory.[36]

He went on," *let me wish your union every success in its championship of working class rights and interests in working for world peace. To many members of your union, I would like to say I wish them every success in their work and every happiness in their lives*".[37] He concluded with," *I was a worker too, a foundry worker for a long time. Although I am doing a different job now, I am still a foundry worker at heart*".[38]

Fred Hollingsworth, a former moulder himself, concluded that the "*greatest significance for us is that Yuri Gagarin belongs to the working class*".[39] The ceremony was closed in the familiar trade union tradition of singing "*For he's a jolly good fellow*". In addition to other gifts, the General Secretary of the AUFW Jim Gardner presented Gagarin with copies of two novels he had written, *No Time for Sleeping* and *He Must so Live*. It is unlikely that these novels were in Russian, so Gagarin could never have read them.

The recollections of one of the youngest members amongst the spectators, and thus probably some of the most tenuous, are from Liam Grundy. Along with his mother and grandfather, at the age of only four years, he was in the crowd outside the AUFW headquarters. Liam, now a musician, remembers the event but not any of the detail, because he had to go into hospital for treatment on the following day, so he associates this visit with what was his "*last day of freedom*".[40]

Also getting wet in the crowd was a then thirty-three-year-old Stanislava Sajawicz. Now an octogenarian, she remembers the day that Gagarin arrived outside the union office with a smile.[41] Her journey to Britain from her birth country Poland sounds remarkable now, but is probably a story of displacement common in time of war. Separated from the rest of her family

4. Together Moulding a Better World

aged twelve, she had been forcibly transported to a work camp in Oblast region of Russia in 1940. Whilst ill, she was befriended by a female doctor and finally left Russia on a troop transporter ship arriving in Tehran in 1942. After staying in temporary homes in Egypt and Palestine, she eventually arrived in Liverpool in 1947. She met her husband Jan in Whitchurch in Shropshire, and immediately after their wedding in 1951, they settled in Manchester where there was a growing Polish community. It was in her local Polish club in Old Trafford that she became aware of Gagarin's planned visit.

She recalls the excitement surrounding the visit even though the weather was not as nice as had been forecast and she had to wait for up to two hours: "*I remember him coming around the corner and he got out of the car, he looked around and I noticed he had a beautiful face. He was a very handsome man, that's what I noticed first. Everything was so nice and peaceful.*" On reflecting on her own and her family's treatment at the hands of the Soviet government, she insisted that she had no resentment for the Russian people. "*I never criticised the Russian people. Russian people are fantastic people. It was not the Russian people but the government people who had been responsible during the war. Russian people are very, very warm people who experienced extreme hardship during the war themselves*".[42]

Although a speaker of Russian, she did not try to speak to Gagarin as he got out of the car because she says she *"was shy"*. Besides, because of the noise and the large number of people he probably would not have heard. With a subtle smile and a narrowing of her eyes, she says, "*I wish I could have said 'welcome' in Russian to him as he went in*". She along with many others did not wait for him to come out of the union office and left as soon as he entered. This meeting with Gagarin left a lasting impression: "*It was raining when he arrived actually, but when I recall the event, I don't remember the rain.*" In subsequent conversations about Gagarin's visit between her Polish friends, a common theme appeared. They felt sorry that such a beautiful person had to live and work in a repressive regime that did not provide a good quality of life to most of its citizens.

In 1986, the union moved from the building at Old Trafford that Gagarin had visited to a new office in Prestwich in north Manchester. During the move, an unlabelled super eight cine film was discovered in a cupboard by Alf Lloyd the Regional Officer of the then Amalgamated Engineering and Electrical Union (AEEU).[43] Along with his colleague Brian Salt, he located a cine projector and previewed the film, which was the only way to discover its content. The film, probably the combined work of several cameramen, was a record of Gagarin's arrival at Manchester Ringway Airport, the medal presentation ceremony at AUFW headquarters (with some audio), Gagarin's tour of the Metropolitan Vickers factory in Trafford Park, and his visit to Manchester Town Hall, also with some audio.

This film is the only recording of Gagarin speaking, in Russian, during his visit to Manchester. Alf Lloyd recalls that the film was almost discarded along with other unwanted equipment, as is the norm at times of office moves.[44] In 2001, Alf Lloyd contacted Professor Jim Aulich, from the Manchester Metropolitan University, who at that time was organising a poster exhibition to mark the 40th anniversary of Gagarin's visit. Aulich recommended the film passed on to the North-West Film Archive where it currently resides.[45]

5. Working Class Cosmonaut

Of all places to visit in Britain, Manchester, the cradle of the Industrial Revolution and its working class and trade union traditions, was perhaps the most apt for the cosmonaut to reconnect with his own working-class roots. The prospect of meeting working men and women in an industrial setting was the one venue of his choosing firmly on the itinerary even before he left the Soviet Union.

At the beginning of the nineteenth century, Manchester had a population of less than a third of one million. Initially through the rapid growth in the cotton industry, the transport network, the development of heavy engineering and the leading-edge technology in machine manufacture, the population grew to over two million by the beginning of the twentieth century. Much of the production from the factories was destined for export, and the key element in the transport chain was the port of Liverpool 57 km away.

Liverpool was the door through which goods manufactured in Manchester found their way to the countries of the empire and especially via the Atlantic to America. It seemed to the hard-working Mancunians that they had to toil in the foundries, factories and mills for their wealth whilst Liverpool could simply take advantage of its geography. As the capacity for manufacturing grew, so did international trade and the demand on the port of Liverpool. Industrialists in Manchester had to pay for transport of goods to Liverpool and then pay the tolls and harbour fees, controlled solely by Liverpool.

In the early nineteenth century, a substantial percentage of the world's trade used the port of Liverpool and the city's wealth grew. Liverpool's port charges remained suspiciously high. The transport costs of sending finished cotton goods to India exceeded the price of the goods themselves.[1]

6. A Delightful Fellow

By the mid-nineteenth century, although the growing railway network eased cargo transport and narrow boats still using the Bridgewater Canal, the cost of export via Liverpool was becoming expensive for the Manchester-based businessmen. In their mutual pursuit for profit, a standoff arose between Manchester and Liverpool. The recession of the 1870s was finally the motivation to kick-start a project that had been suggested several years earlier.

The ever-expanding manufacturing base in heavy and precision engineering that powered Manchester's growth enabled industrialists to take on a radical new project to end the city's land-locked status. The 57-km long Manchester Ship Canal took over 16,000 men thirteen years to build using earth dredgers, steam excavators, steam cranes and locomotives. It was a monumental undertaking that included five sets of locks to raise ships up to 20m and seven swing road bridges. The first and only swing aqueduct in the world is at Barton. Built in 1893, it carries the Bridgewater Canal over the Ship Canal and is still in service today.[2] The Manchester Ship Canal was first used on 1st of January 1894. A diamond-shaped piece of land twelve square kilometres in size became the world's first planned industrial estate. Bounded by the Manchester Ship Canal to the north and the Bridgewater Canal to the south, it was conveniently located just 5.5 km south-west of Manchester city centre. Between 1911 and 1918 it had Manchester's first purpose built airfield.

It was not successful for the first few years, but then in 1899 the American George Westinghouse acquired a site in Trafford Park and established the British Westinghouse Electrical and Manufacturing Company. The site was ideal for companies that dealt with large, bulky or heavy products and needed a direct link to America. An inventor and experienced businessman, Westinghouse already had several factories in America and he introduced the streamlined industrial production methods to

6. A Delightful Fellow

Trafford Park by sending his English managers, supervisors and factory foremen to his factories in Pittsburgh for training. Decisions taken in the company's main office block, locally known as "The Big House" and modelled on an existing building in Pittsburgh, shaped Trafford Park's destiny over decades, until it was finally demolished in the 1990s. Westinghouse also built around seven hundred dwellings, schools and shops for his employees, planned in the American style with avenues north and south, streets east and west.

Figure 13 Yuri Gagarin arriving at Metropolitan Vickers
(Courtesy Marx Memorial Library)

In 1919, British Westinghouse Electrical and Manufacturing Company merged with Metropolitan Vickers Electrical Company. Following further mergers, a decade later it became Associated Electrical Industries or AEI. This was the formal company name at the time of Gagarin's visit, but locally it was still popularly known as Metropolitan Vickers or shortened to Metrovicks. Not much of George Westinghouse's site exists today, but he is remembered by a key thoroughfare, Westinghouse Road, that cuts through the centre of Trafford Park.

6. A Delightful Fellow

Throughout its history, the large complex of Metrovicks, including its foundries, has been associated with the production of high quality, heavy-duty industrial machines for internal use, but especially for export: particularly transformers, generators, turbines, dynamos, high voltage circuit breakers, traction engines, turbo generators and electric locomotives. Power stations, industrial plants and transport systems were supplied by Metrovicks to countries around the world including China, India, South Africa, Egypt and the Soviet Union. During the war, Metrovicks had supported the war effort producing munitions, aircraft, radio transmitters and receivers. After the war, Metrovicks fell into decline: a site with some of the oldest and largest foundries that had once employed over 23,000 people was by the mid-1990s left with a turbine refurbish plant employing only around a hundred workers; it finally closed in 1997.

Gagarin's first-hand experience of foundry work began in 1949 when at the age of fifteen he left home for the first time to study foundry work in the vocational school in Lyubertsy. His initial impression of a foundry was a scary one, *"Everywhere there was fire, smoke, steams of molten metal and workers in overalls busy at their jobs"*.[3] When along with his classmates he was shown the foundry for the first time, the foundry foreman welcomed them with the words *"Take a good look, get used to handling fire. Fire is strong, water is stronger than fire, Earth is stronger than water but Man is the strongest of all"*.[4] The fear did not persist and soon he was learning the skills of a moulder, creating moulds with cores and moulding sand. The course involved theoretical as well as practical work.

In his highly censored autobiography, *Road to the Stars*, Gagarin recalls his first pay-day. It was not a substantial amount but it was the first money he had ever earned. He sent half to his parents. Much of the book is overt communist propaganda but the description of his devotion to his family and empathy with working men and women carries an element of sincerity. Six

6. A Delightful Fellow

years later, he would experience the lavish lifestyle and the attention of an international celebrity and be guest of kings, queens, prime ministers and presidents, but up until the age of twenty-one, a working-class lifestyle was the only life he had known. Perhaps it was the experience in those formative years that almost a decade later drew him to accept an invitation from a foundry workers' union in a faraway city known for championing the rights of the working-class people.

Prior to his single orbit of the Earth, he had never left the Soviet Union. It is however possible that he knew about the industrial city of Manchester in the north west of England before he received his invitation to visit it. After all, much of its textile and industrial produce was destined for export. In the 19th century, Manchester was known as Cottonopolis (cotton metropolis). During its heydey in 1853, it had 108 steam and water powered mills.[5] Even today in the twenty-first century, long after its international eminence in textiles production ceased, the Australian term for cotton bed sheets and pillow cases is "Manchester", and in some parts of Germany corduroy trousers are known as "Manchesterhosen".

The Soviet education system would have featured the works of Karl Marx and Friedrich Engels of which Gagarin would have been aware. Friedrich Engels wrote the classic *"Conditions of the Working Class in England 1844"* whilst living in Manchester. In that work Engels documented the appalling conditions of the working classes of Manchester and surrounding towns such as Oldham, Stockport and Bolton. Engels documented the high mortality and dangerous working conditions introduced by the manufacturing industry. He wrote the industrial revolution was" making *the workers machines pure and simple, taking from them the last trace of independent activity"*.[6] Gagarin was extremely well read, he would probably have been familiar with Manchester and its working-class traditions long before he arrived.

6. A Delightful Fellow

Although the rain had eased as Gagarin left the Amalgamated Union of Foundry Workers building in Old Trafford at around 11.30 a.m., the crowds along the road had not. His route to Metrovicks would take him past the Manchester United football ground. The players interrupted their training and came out to join the cheering crowds on the roadside.

Figure 14 Yuri Gagarin inside Metropolitan Vickers (Courtesy Alf Lloyd)

Gagarin's Bentley had driven along Westinghouse Road, dry but the occasional puddle was evidence of the sharp showers earlier. He turned right under Northgate facing Third Avenue and entered the large Metrovicks complex where the car stopped, and he started his tour on foot. Surrounded by photographers, union officials, his interpreter and some senior staff from the factory he signed the visitors' book and proceeded towards the foundry to meet the foundry workers on the shop floor.

It was there that Stanley Nelson was introduced to him. Despite knowing that he did not understand English, Nelson spontaneously said *"Hello and welcome"* as the spaceman was quickly moved on. The AUFW had independently invited Gagarin to Manchester so they asserted their control over his

6. A Delightful Fellow

itinerary. Working as a draughtsman in the design office of Metropolitan Vickers in Trafford Park located close to the foundry, Nelson had not been aware of the planned visit by the cosmonaut. He was surprised by the haste by which Gagarin was whisked through the heart of the Metrovicks plant, and by the strong controlling influence of the union men who remained in close proximity and managed where Gagarin went and whom he met. He recalls that it was, *"very much a union controlled event. He was closely managed by union minders throughout his visit. The Metrovicks management played no part, only the unions"*.[7]

Students in schools and colleges, and young apprentices in industry, were supported by the Metrovicks education department where Stanley Nelson's wife Joyce worked. She was on site that day but did not see Gagarin. During the fifties, she routinely walked past the foundries like Taylor Brothers and Turner Brothers in Trafford Park where the doors were left open to aid temperature control and ventilation. For anyone peering in, she remembers, *"it was like a vision of hell. Smoke, fire and tiny thin men silhouetted against the foundry fire. No one was fat; they were all thin like Lowry's match stick men"*.[8] Gagarin had survived the multiple unknown dangers of the new environment of space in his record-breaking spaceflight. Seeing the foundry conditions at first hand after such a long time would have reminded him that foundry work was not without its own danger.

Gagarin would have been familiar with the foundry layout but surprised by the large scale of the complex, which was probably the largest he ever visited. Groups of men downed tools and greeted the visitor mainly with a smile, cheer and wave, or for those close enough, a firm handshake. In contrast to the opulent surroundings of state visits in Hungary and Czechoslovakia and the impending visit to Buckingham Palace, here he was surrounded by the smell, smoke and dirt of a working foundry and foundrymen in their work clothes, with the light-attenuating goggles with which they peered deep inside the furnaces

6. A Delightful Fellow

hanging around their necks. Some with soot on their faces stretched out rough, dirty strong hands that Gagarin gladly shook. He appeared genuinely at home despite the language barrier and comfortable in an environment he understood well. At a time when he was still the only person in the world who had orbited the Earth, for a short while he was taken back to his foundry worker days. A photographer stood on a cast to get a better picture as Gagarin walked by. Gagarin shook his head with disapproval and through his interpreter indicated that the cast was not complete and he should not be standing on it.[9]

The minders kept Gagarin on the move and this prevented a repetition of the congestion that had brought his tour of the Soviet Trade Fair at Earl's Court to a standstill on the previous day. He walked back to the main aisle and then crossed into the manufacturing aisle where casts from the foundry were machined and assembled into final components.

Conveniently, it was now around midday and, for many, time for a lunch break. The final item on Gagarin's Trafford Park itinerary was an informal speech in a car park, known locally as Cinder Car Park. It was a venue familiar to all the shop floor workers as a meeting place, especially at times of industrial unrest. To get there, Gagarin had to come back from the manufacturing aisles and walk back along the main aisle towards Northgate, passing on his left the redbrick office block, the "Big House" that George Westinghouse had built, an exact copy of his headquarters in Pittsburgh.

The AUFW estimated more than ten thousand (the *Daily Telegraph* on 13th July 1961 estimated half that) were present in the car park to hear Gagarin make a short speech from the back of a lorry. It was probably the most eclectic crowd that had convened in that car park. In addition to the cloth-capped workers and suited clerical and management staff, there were many young women and children, too. The car park was publicly accessible, so many present were not local workers but visitors

6. A Delightful Fellow

for the day. Many standing near the front of the lorry that formed Gagarin's platform reached up and stole a handshake. Gagarin's handsome looks were acknowledged universally but it was the women who expressed it openly.

Figure 15 The "Big House" (courtesy Stanley Nelson)

As the women at the airport and the union office had done, the women in the car park commented on his good looks, "*Oh isn't he nice*", whilst one seasoned male worker noted Gagarin's stature, "*he's only a bit of a lad*", and another, uncomfortable with speeches, suggested that "*it would be better having a pint with the lad*". The lorry was prepared with two microphones, one for Gagarin and the other for his translator. Following a short introduction by the shop steward, Bert Brennan, Gagarin spoke, his bright green uniform in stark contrast to the grey and black work clothes of his audience. Accompanied by his translator Belitzky on the lorry, he empathised with the large workforce gathered in front of him. This former foundry worker had insisted at the union office that he was still a foundry worker at heart. He spoke of the thousands of Soviet scientists, engineers and designers who had collectively made his flight possible. He told them that following his flight the Soviet government had made 7000 awards to those closely involved in his successful mission, before returning to the theme of peace, friendship and collaboration. He said, *"There is room for all in outer space. There is room enough for Russians, Americans and for the British as well as many other people"*.[10]

Firm Handshakes

Gagarin's speech was received well, although most either did not hear or did not understand what he had said. One of the outstretched hands that he shook from the lorry belonged to Edith Hope. She had waited in the rain long before Gagarin arrived to claim her place near the front. At the age of 71, she was one of the oldest in the car park that day. During the First World War she had been a labourer to a moulder at the same foundry in Trafford Park where her husband had worked and her son was working at the time. Her working day had consisted mostly of transporting heavy loads by wheelbarrow from one part of the foundry to another.

Working conditions for unskilled workers are traditionally demanding, but even more so for a slim twenty-five year old mother of three; she recalls in a letter to the union after Gagarin's visit,[11] that one *"big head"* forced her to fill her wheelbarrow so much that she could not physically move it. In the absence of a union, she had no one to turn to, *"If only your union was as strong then as it is now"*, she wrote. Having seen a picture of her handshake with the cosmonaut that her son had discovered, she shared her sentiment in a letter to the union, *"I was repaid with a humble boy bending down to take my hand"*, a sentiment which was repeated in spirit throughout the day by thousands of others. The union had sent the invitation, but it was Gagarin's own desire to make a personal connection with the

6. A Delightful Fellow

working people to whom he had belonged only six years ago.

Through this brief and unique shared experience amongst the industrial grime of a large manufacturing workplace, one charismatic Soviet man, a former foundry worker, provided the people of the west with an insight into the people of the east. The experience of the encounter with the workers in Trafford Park left a deep impression on the spaceman. On the anniversary of his spaceflight, 12th April 1962, Gagarin sent a radio message via Moscow Radio to the AUFW in which he said, *"The firm handshakes of my fellow workers in the moulding workshop were dearer to me than many awards"*.[12] Following the brief speech, it was time for his last venue – Manchester Town Hall.

Figure 16 Yuri Gagarin at the Cenotaph, London
(Courtesy RIA Novosti)

Keith Fairhurst, a young gas distribution engineer working at the old Gaythorn site in Medlock Street leading to the city centre, now demolished, had been allowed with other workmates to see the cosmonaut. He remembers *"the man himself sitting high in the back of the open car, smiling and waving to us. I recall that we all noted how small he was and the large, peaked, Russian uniform cap he was wearing. He went slowly past, and then it was back to work."*[13]

Medlock Street leads directly to St. Peter's Square, the site of a Portland stone cenotaph designed by Sir Edwin Lutyens. It was

6. A Delightful Fellow

erected in 1924 to honour the memory of those who gave their lives for their country, initially commemorating the great loss of the First World War, but now both wars. Gagarin did not visit this cenotaph but made a brief stop to lay a wreath during his time in London.

Despite the police outriders, Gagarin was once again delayed by strong-willed masses that lined the approach to Albert Square, the large open area in front of the Town Hall entrance. With assistance from mounted police, his car eventually drove up to Albert Square and stopped outside the Town Hall entrance, chased by a couple of energetic photographers on foot. The rain had ceased, and the boundless enthusiasm of the Mancunians outside the Town Hall seemed to exceed even that of the crowds in Old Trafford and Trafford Park. It was about one o'clock now, and the large open square was ideally located in the city centre for workers from the offices and shops to take a convenient lunch break to welcome the cosmonaut.

Ray Smith, working as a calligrapher on the top floor of the office block in Albert Square opposite the Town Hall, had a grandstand view as Gagarin was greeted on the steps of the Town Hall draped by a Soviet flag with the Soviet national anthem playing in the background. Ray estimates the crowds to have been at least 3000, many chanting the cosmonaut's name. From his vantage point, he had seen many VIPs in the past, but never forgotten that day which inspired his passion for astronomy that is still firing him fifty years on.[14] Armed with his black and white camera, 33-year old Fred Ritchie had taken time off work from a city centre textile company to be amongst the crowds at Albert Square outside the Town Hall for Gagarin's arrival. His colleagues did not share his passion and chose not to make the short journey to Albert Square. Fred had waited half an hour before Gagarin's car arrived and still cherishes the picture he took that day.

Gagarin entered the neogothic Grade 1 listed Town Hall and

6. A Delightful Fellow

once again met the Lord Mayor who had welcomed him earlier in the rain on the tarmac at Manchester Airport but now greeted him as the guest of honour to the civic reception. Following the formal introductions and signing the visitor's book, the hosts and their guest ascended the curved staircase to the first floor onto a reception area covered by a curved stained-glass roof listing the names of all the former lord mayors and the floor tiled with images of a bee.[15]

Figure 17 Yuri Gagarin arriving at Manchester Town Hall
(Courtesy Fred Ritchie)

The textile industry had been the source of Manchester's wealth that funded the splendid Victorian Gothic building. Built with hard sandstone quarried from nearby Bradford, the Town Hall looked older than its mere 84 years at the time of Gagarin's visit. Perhaps that was the intention of its architect Alfred Waterhouse, who also designed the nearby Manchester Museum, Rochdale Town Hall and the Natural History Museum in London. They all share common features of tall towers and grand layered archway entrances.

Gagarin entered a large room decorated with wood panelling, two large chandeliers and three almost floor to ceiling arched windows adorned with intricate stained glass that overlook Albert Square where the crowds were now beginning to diminish. One long table ran along the length of the room, at the centre of which the guest of honour, the Soviet Ambassador, Manchester's Lord Mayor and other dignitaries would sit with their backs to the windows. At right angle to that long table were three shorter ones to accommodate the remaining guests. Bernard Lovell was seated at one of the three tables close to Gagarin.

It was now lunchtime, and this was Gagarin's first proper opportunity to sit down and eat since his arrival at Ringway four hours earlier. The time at the AUFW was too short and the environment too cramped to take advantage of the light meals they had provided. The menu at the Town Hall was something he had probably not come across before. It consisted of five courses: fruit cocktail, sole plaice palace, farm-bred (free range) roast chicken with trimmings, strawberries with cream, and cheese and coffee as the last course. Head chef Michael Byrne assisted by Joe Marceau prepared the menu and it was served by waiters headed by Frank Gerity.[16]

Jodrell Bank

Unlike Sputnik,[17] launched on 4th October 1957, Gagarin's spacecraft Vostok did not fly over Britain. Many radio amateurs in Britain and around the world had listened to the simple but unique radio transmission from the world's first satellite that is now an icon of the start of the space age. The radio frequencies were published by the Soviets, and sophisticated radio equipment was not required to listen to its now familiar beep-beep signal. Many even saw it. Sputnik itself was just on the

6. A Delightful Fellow

edge of human naked eye visibility, but what most people actually saw was the spent second stage carrier rocket which was much bigger and brighter than the small beach ball sized Sputnik. The carrier rocket had sufficient momentum to reach an almost identical Earth orbit, with Sputnik only a short distance ahead. The radio operators, on the other hand, were troubled by neither daylight nor bad weather, so they could monitor Sputnik's daytime transit over the British sky, too.

The launch of Sputnik was timely for Sir Bernard Lovell, who had led the project to build the Mark One Telescope (now called the Lovell Telescope) at Jodrell Bank, 30 km south of Manchester. The seventy-six-metre diameter dish was then the world's largest fully steerable radio telescope. Jodrell Bank was involved with many of the Soviet space missions in the 1960s. Bernard Lovell would eventually get used to the overwhelming demands of the media but he was first introduced to it following the launch of Sputnik on Friday 4th October, which triggered a "*state of siege of newspaper and broadcasting personnel*"[18] over the weekend at Jodrell Bank. Signals from Sputnik could be received by simple radio receivers, which had already been done, but the press wanted to hear the signals received via the new large telescope. It would have been possible, but Lovell refused to engage in this "*pointless exercise*".[19]

Lovell had already been working on using the telescope as a radar device for astronomical research. By transmitting radio waves to the Moon and then receiving the echoes on the same dish, it was possible to investigate the nature of the material in space between the Earth and the Moon. A week later, on Saturday 12th October, using this radar technique, Jodrell Bank detected the second stage of the rocket that had launched Sputnik as it moved over the British Lake District towards the North Sea at eight kilometres a second.[20] At that time, this was the only telescope in Europe that could have accomplished this feat. Lovell emphasised the significance of this achievement by reminding the press that this rocket was by design a Russian

6. A Delightful Fellow

Intercontinental Ballistic Missile, reinforcing the threat that Russian technology posed to its Cold War opponents.

There were ninety guests from business, commerce and trade unions at the hastily arranged reception that included the director of Jodrell Bank – Sir Bernard Lovell. The Soviets had invited Lovell to join their space effort even before Gagarin's selection as a cosmonaut in December 1959. Lovell had been involved in tracking Soviet satellites since Sputnik in October 1957, but he had also been approached by the Soviets to assist in detecting signals from an ambitious lunar probe (later called Luna 3) in October 1959.

Luna 3 was launched on 4th October 1959, the second anniversary of Sputnik, into a very unusual elongated Earth orbit which stretched out past the Moon. On 7th October, it passed the Moon at 40,000 km and took twenty-nine pictures of the far side of the Moon in forty minutes. No one in human history had ever seen the other side of the Moon before.

During the planning phase for Luna 3, an option to take pictures of the Earth from the vicinity of the Moon was considered but rejected, clearly a missed opportunity. It is in the nature of exploring the unknown that it is unclear at the time which discoveries eventually turn out to be profound or trivial. Almost a decade later, during the first manned flight to the Moon of Apollo 8 in 1968, the first humans experienced a view of the Earth from the Moon that became an icon of the space age. As William Anders, crew member of Apollo 8, put it: *"After all the training and studying we'd done as pilots and engineers to get to the moon safely and get back, [and] as human beings to explore moon orbit, what we really discovered was the planet Earth."*[21]

The mission was one of the remarkable Soviet successes based on exquisite technical innovation in design, manufacture and execution. Digital photography was some way off. After the pictures were taken by the on-board camera when it arrived at

6. A Delightful Fellow

the other side of the Moon, they were developed, fixed, washed and dried within the onboard dark room, then scanned and transmitted as an analogue radio signal when Luna 3 arrived in the vicinity of the Earth 11 days later. This was one of several instances where the large steerable radio telescope at Jodrell Bank was uniquely placed to assist in the Soviet space programme by detecting weak signals from a fast-moving small object. Amongst other instances of support, it was for the assistance with Luna 3 that Gagarin wanted to thank Lovell.

Through his own experience inside the Soviet space programme, Gagarin is very likely to have known about Jodrell Bank, especially its location, and perhaps that was another factor in his consideration when he accepted the invitation. Now at the age of 97, Lovell has no comment on whether Gagarin's choice in coming to Manchester was related in any way to the nearby location of Jodrell Bank.

During the formal speeches at the Town Hall, guided possibly by instructions from his seniors but certainly out of professional courtesy, Gagarin paid the following tribute to the work done by Lovell and his colleagues at Jodrell Bank.

"It has given us special pleasure to meet so eminent a scientist as Sir Bernard Lovell. He has given great help to our scientists in the tracking of satellites and space rockets and we regard this as a great example of scientific co-operation in the peaceful exploration of space. We trust that these contacts will thrive and develop for the peaceful exploration of outer space for the benefit of mankind's scientific progress".

In a press interview, Lovell lamented that he would have liked to ask Gagarin some technical questions but it was unlikely that he would have got answer. Lovell had personally invited Gagarin to Jodrell Bank Telescope 30 km away but Gagarin's schedule would not permit it.

6. A Delightful Fellow

Francis French is a successful author specialising in space history now based in California. During the seventies and eighties, he went to school in Manchester, and as a space fan from an early age he was very familiar with Jodrell Bank. In 1987, whilst researching Gagarin's visit to Manchester, to his surprise he ended up speaking to Lovell about the cosmonaut's visit. French recalls:[22]

"I called Jodrell Bank and asked if they had any information or photos of Lovell meeting Gagarin. Having read that Lovell had retired in 1981, I was very surprised when I was told Lovell was there that day, and they put my call through to him. Somewhat awestruck, I had an enjoyable conversation with him. Like everyone Gagarin met on that trip, it seems, Lovell was struck by Gagarin's warm smile and star quality... The large numbers of people hoping to greet Gagarin even in the relative protection of Manchester's town hall meant Lovell didn't even try to ask him technical questions. He also knew that Gagarin, a master of answering questions without revealing any secrets, would not be allowed to tell him. Despite the lack of personal time with the cosmonaut, Lovell was thoroughly charmed by him."

Despite the fractured east-west relationship during this period of the Cold War, there were other instances of collaboration. The Soviets provided Patrick Moore with copies of Luna 3 images that arrived at the BBC five minutes before the live transmission of *Sky at Night* in October 1959. Moore had been observing the Moon for many years and was considered an expert in lunar mapping. He shared his knowledge with both the American and Soviet space programmes, so it was no surprise that he was granted access to them so soon after their reception on Earth. Even with only so few images returned, the mission was considered as an astonishing success, by those in and outside the Soviet Union.

In the AUFW cine film of the Town Hall visit, Gagarin returned to the topic of the Soviet attitude to further international

6. A Delightful Fellow

collaboration in space: *"if the USSR combined its efforts with the entire west we could make even faster progress in the exploration of space and have even greater achievements to our credit."* When asked about his own plans for further spaceflight, he replied through his interpreter, *"In Russia we have a saying, the appetite grows with eating. Now having made my first flight and enjoyed it, I would like naturally like to fly to the Moon then perhaps to Mars and Venus and even further if my abilities make it possible"*.

Figure 18 Yuri Gagarin at Manchester Town Hall (Courtesy Alf Lloyd)

The aircraft that brought Gagarin to Manchester waited on the tarmac to take him back to London. As the reception drew to a close, many, including Sir Bernard Lovell, approached him for a personal comment. The crowd outside had waned, but still, in the middle of a large city during a working day, there were many onlookers, some who happened to see him by chance, only discovering the identity of the man in the green military uniform once he had driven by. It was now about 3.30, still inside working hours, still too early for many office workers to walk out to the street. Instead, doorways, balconies and windows with a view were occupied by curious office workers.

Brenda Nowell was 27 years old when she came across the cosmonaut – who was also 27 years old. At the time, she was

6. A Delightful Fellow

working for National Employers, an insurance company around the corner from the Town Hall, and saw Gagarin from the first-floor office on Princess Street as his car turned right from Albert Square as he left the Town Hall.

It was now warm and sunny and people were still cheering and clapping, and some were throwing flowers into the open top Bentley. Brenda remembers, *"He looked very relaxed and confident, with his friendly smile he waved at us"*.[23] Gagarin was still in his uniform, still waving, still smiling the tenacious smile with which he started the day. She treasures the memory and a photograph taken by her colleague Frederick Nash, and has repeatedly told the story to her grandchildren.

Figure 19 Yuri Gagarin on Princess Street returning to the airport
(Courtesy Brenda Nowell)

Gagarin drove along Upper Brook Street away from the Town Hall, and although the dense crowds had now gone, he remained standing in his car, waving back to those queuing along the pavement, in doorways and looking down from their first-floor bedrooms. On his way to Princess Parkway he passed along Wilbraham Road.

Heavily pregnant, Patricia Patton had brought her two sons to

6. A Delightful Fellow

Platt Fields Park and joined in with the crowds congregating on both sides of Wilbraham Road. She has kept a copy of the *Evening Chronicle* front page – slightly faded now – with the headline "Man in Space Back Alive". Almost fifty years on, she still has a strong memory, although she remembers Gagarin wearing a brown rather than a green uniform. In her winning entry to a writing competition in 2009, she wrote:

"We could see him standing erect, very handsome, wearing an immaculate brown uniform, hand stiffly raised in the form of a salute – a god, a film star – an alien! It was a unique experience and we all cheered and clapped with all our hearts. Today we are bombarded with a celebrity culture of little significance, so apart from, say, Nelson Mandela or the Queen, I don't think now I would ever stand and wait, as I did with such excitement to see a true hero."[24]

Figure 20 Yuri Gagarin leaving Manchester. Upper Brook Street Manchester (Courtesy Gareth Khodna)

Gagarin arrived at Ringway Airport, dry but half an hour later than the planned 4:00 p.m. The aircraft that had brought him and its crew would now take him back to London. Fred Hollingsworth, the president of the AUFW, was one of the last to shake Gagarin's hand as he boarded the aircraft. The Major

6. A Delightful Fellow

ascended the stairs and looked back and waved one last time to the people of Manchester before entering the aircraft. The union executive council and other members of the union standing nearby on the tarmac waved back. Fred Hollingsworth was elated by the overwhelmingly successful day for him and his union, later recalling, "*This has proved to be the greatest day in the history of our union*".[25]

This was Gagarin's second day in Britain and invitations from the Prime Minister and Buckingham Palace had now been received and accepted. It is conceivable that Gagarin originally accepted the invitation from the Manchester-based AUFW with the intention of visiting the Jodrell Bank Radio Observatory too, but events simply overtook him. Despite initial efforts of Her Majesty's Government to ensure this was not an official state visit it was now turning into one.

6. A Delightful Fellow

Gagarin's third day in Britain started once more with heavy rain. The mid-summer weather would swing between heavy sudden downpours and a hot summer's day. Back in his Rolls Royce with the personalised number plate YG1, he left his base, the Soviet Embassy in Kensington, at a leisurely 10:00 a.m. for his first engagement at Lord Mayor of London' residence, Mansion House at 11:00 a.m. The venue was only 8 km away, but having missed out on a day-time tour of the city on the day he arrived he would take a longer route to experience some of the sights of London before his first engagement.

This third day's schedule was varied. It included a formal visit to Mansion House, an informal visit to the Tower of London, a meeting in the learned environment of the Royal Society, and, after a break back at the Soviet Embassy for a late lunch, the most significant appointment so far, a meeting at the Admiralty with the British Prime Minister. A unique one-to-one meeting, also at the Admiralty, with a fellow test pilot, followed that. The day's programme concluded with a further two receptions in the evening.

His route from the Soviet Embassy to Mansion House was designed to give him a brief tour of the city that was his home for five days. From Kensington it is a five-minute drive to Mansion House on the other side of Hyde Park. The scenic route took him past Buckingham Palace and the Houses of Parliament, past Lambeth Palace, back over Westminster Bridge, Trafalgar Square, the Strand and Fleet Street, and ended conveniently outside Mansion House punctually an hour later. As he toured the city it rained, the car roof remained down, and the overcast dull day conspired to give him a disappointing but very "British" experience.

6. A Delightful Fellow

Mansion House is the official residence of the Mayor of London. This visit was added at short notice, a consequence of the additional time he now had in Britain resulting from the extended stay following the invitations from the Prime Minister and then the Queen.

Following the noise, smell and dirt of a working foundry in Manchester, Mansion House with its fluted Corinthian columns, spectacular interior and elegant furniture could not have been more of a contrast. It is a magnificent three-story edifice with the air of a Greek temple, uncomfortably squeezed amongst the functional architecture of the surrounding mostly twentieth century buildings. Built in the middle of the eighteenth century on the site of a former church destroyed by the Great Fire of London, it continues to be a venue for important city functions.

During its early history, Mansion House dispensed justice with a purpose-built court and cells in the basement. In 1920, Sylvia Pankhurst, daughter of the Manchester-born suffragist Emmeline Pankhurst, convicted for publishing articles considered seditious, was imprisoned here for six months in the basement. The cells have long since been converted to a wine cellar, and the courtroom is now the Lord Mayor's international and domestic office. Sir Bernard Nathaniel Waley-Cohen had become London's Lord Mayor in 1960, so the duty of hosting the spaceman's visit fell to him. Waley-Cohen was a successful businessman with interests in real estate, investment banking and farming. As director of the Palestine Corporation, he had assisted the development of Israel soon after it was founded. On the day of Gagarin's visit he was adorned in the regalia of his office, and along with his deputy he wore a black suit, white frilled shirt and his polished ceremonial chains of office.

Gagarin arrived in his Rolls Royce with the roof still closed. A few thousand spectators had gathered over the previous hour, huddled under their umbrellas as they lined the pavement along

6. A Delightful Fellow

Queen Street and Cannon Street. Cameramen along with the spectators got no respite from the rain during this brief visit. The scaffolding erected along one side of Mansion House long before the invitation of the spaceman did not interfere with the view. The mayor welcomed him with the words, *"People of Britain salute Yuri Gagarin. We are thrilled that he has found time to visit our country"*. Gagarin replied that *"he had been impressed by London's greenery, grandeur and architecture"*.[26] In Manchester he had felt connected to the people and places. Here in the capital he seemed to be going through the motions of fulfilling a formal obligation.

Inside the grand building with the backdrop of its fine décor, exquisite chandeliers and unique art collection, Gagarin made a short speech, again simultaneously translated by Boris Belitzky. It was a short symbolic thirty-minute visit like so many others. Before leaving, Gagarin along with the mayor and other VIPs came out to the elevated entrance under the pediment supported by six mighty Corinthian columns and waved to the crowd but did not address them. At Mansion House, he kept at a distance from the admiring crowd. That would not be possible at his next venue.

The next stop was a 1 km drive east along the northern bank of the River Thames close to Tower Bridge. The Tower of London is one of the most popular London attractions for tourists. It is a collection of buildings arranged in two concentric circles dating back to the eleventh century. In its time, it has served as a castle, treasury, royal mint, prison and a residential royal palace. Its major attraction today is the crown jewels that consist of an assortment of crowns, rings, sceptres, orbs, robes and more. The collection includes the fabled Koh-i-Noor diamond in the Imperial State Crown. Some of the jewels are worn only during the coronation ceremony and formal state functions, so unlike the Lord Mayors of London and Manchester, the Queen would not be wearing any of these ceremonial decorations during her meeting with the spaceman for lunch on the following day. The

6. A Delightful Fellow

Communist spaceman's reaction to these symbols of royalty is not recorded.

Figure 21 Outside the Tower of London (Courtesy Marx Memorial Library)

Mid-July is the middle of the tourist season for London. The gridlock that ensued at the Tower of London with Gagarin's attendance was not a surprise. It was an unexpectedly colourful occasion as the spaceman toured the famous site in his khaki top and blue trousers accompanied by the beefeaters in their traditional bright red Tudor dress. In turn, they were surrounded by press, cameramen and the helmeted police outriders who had joined in trying to maintain a path for the whole party.

Donald McCormick was making his first visit back to the UK after having emigrated with his wife and two children to Canada in 1956. After visiting relatives, they had taken a few days out to explore London. During their tour of the historic Tower of London, the McCormicks were surprised by the unexpected melee and commotion and would only learn from others in the crowd that like them the world's first spaceman was touring the Tower.[27]

6. A Delightful Fellow

By the time Gagarin had seen the Crown Jewels, the crowd of tourists and schoolchildren had grown so large that he could not get to his car and, briefly trapped, he had to wait for the car to come to him. Amongst the school parties on that day was a fifteen-year-old Dave Monk on his summer school trip to the Tower of London from Caludon Castle Boys Comprehensive School in Coventry. Although he went on to be a photographer later on in life, he was not quick enough to capture a picture of the cosmonaut on his Brownie 127 during this chance encounter.

Nearly fifty years on he can still clearly recall the cosmonaut's smiling face. Now retired after a career as a tester with General Electrical Company, Monk now does voluntary work with the Coventry Society for the Blind. The memory of seeing Gagarin at the Tower of London is his most vivid schoolboy recollection. As a youngster, he suffered from a speech impediment and this brief encounter gave him added confidence and inspiration to face the world as an adult in the following year. "*By stroke of luck*", he recalls, "*I found myself in a crowd of people outside the Tower trying to catch a glimpse of him. Only a few feet away, I saw this Russian hero wearing a military cap and a friendly smiling face*".[28] This chance meeting also left him with a life long interest in space and led him in 1991 to interview other spacemen: cosmonauts Colonel Alexander Volkov and Dr Alexander Martynov during their visit to Coventry Technical College. Later in 2010, he recorded an interview with Apollo 16 astronaut Charlie Duke for *Coventry's Talking Magazine for the Blind and Partially Sighted.*[29]

Eventually, Gagarin and his cavalcade left the Tower of London and headed back west parallel to the river towards and then through Piccadilly Circus. The convoy then turned into the forecourt at Burlington House and the big iron gates closed behind them.

6. A Delightful Fellow

The Royal Society

The Royal Society is a UK-based fellowship of the world's leading scientists founded in the mid-seventeenth century and is the oldest scientific academy in continuous existence. No formal speeches or photo shoots were planned; this would be an opportunity to eat, mingle and to talk technical with western scientists and engineers. Gagarin was now experienced at the latter without sharing information about what had already been defined as state secrets of Soviet technological innovations. His stern overseer Nikolai Kamanin along with the Soviet Ambassador and others accompanied him, so it was unlikely he would slip up. The language barrier would also help, and if the Soviets had had an Official Secrets Act, the interpreter Belitzky would have been a signatory.

The Royal Society's President Howard Florey, who in 1945 had shared the Nobel Prize for medicine with Ernest Chain and Alexander Fleming for the extraction of penicillin, welcomed Gagarin. Following brief cordial words of welcome and introduction, they all turned to lunch. Gagarin declined the offer of wine and drank only orange juice. To the small group of about twenty-four scientists, Gagarin spoke about the importance of Newton's work on gravitation and the laws of motion that played such a central role in his spaceflight three months earlier. It was inevitable that the conversation would turn to Isaac Newton, who had been not only a Fellow but a former President of the Royal Society.

Bernard Lovell who had been at the Town Hall reception in Manchester on the previous day was also present. Lovell had been elected as a Fellow of the Royal Society in 1955 and despite the long journey from Manchester wanted to be present. Once again, Lovell acknowledged Gagarin's winning personality, saying that he was "*a very even-tempered man, not easily disturbed or worried by anything and completely*

6. A Delightful Fellow

charming."[30] Gagarin spoke once more, as he had done in Manchester Town Hall, of his desire for further international collaboration, adding "*that it should not be restricted to just space exploration but in all scientific fields*".[31]

The President presented the cosmonaut with two volumes of *The Correspondence of Isaac Newton* that Gagarin said he would treasure as a souvenir of this memorable occasion. After lunch Gagarin was shown a few of the library's unique collections, including a copy of the *Principia*, Newton's revolutionary work on gravitation, where he demonstrated how to calculate Johannes Kepler's laws of planetary motion[32] using only mathematics. Kepler himself had discovered the laws empirically by the meticulous examination of records of planetary positions logged by naked eye observations recorded over several decades. Gagarin was also shown Newton's reflecting telescope, the first telescope design to use mirrors not lenses.

Figure 22 Yuri Gagarin in London (Courtesy Marx Memorial Library)

In 1960, the Royal Society had celebrated its three-hundredth anniversary. During that time scientists from around the world had become members. Gagarin was shown certificates of election of Russian members that he may have recognised. They included the famous chemist Dmitri Ivanovich Mendeleev and

6. A Delightful Fellow

physiologist Ivan Petrovich Pavlov. In the light of the honorary awards Gagarin had received from the British Interplanetary Society and the AUFW in Manchester, the idea of a similar, honorary, award may have been briefly considered. The Royal Society did not have a tradition of honorary memberships at the time of Gagarin's visit and still does not. However, starting in 2000, seven[33] Honorary Fellowships of the Royal Society have been awarded to individuals the society considers have rendered special service to the cause of science.

It was now approaching 3:00 p.m. and the meeting at the Admiralty was less than an hour away. It was a 5-km drive from Burlington House to the other side of Hyde Park to the Soviet Embassy. This would be a perfect opportunity for Gagarin to have a pause, gather his thoughts and deal with any last-minute preparation for what was likely to be his most important political engagement of his world tour so far, a meeting with the Prime Minister of Great Britain.

Downing Street was undergoing renovations on 13th July 1961, and under those circumstances, the meeting with the Prime Minister Harold Macmillan would have taken place in the House of Commons. However, that was a camera free zone in those days. An exception to the rule, if the Prime Minister wanted it, was available. A memo to the Prime Minister stated, "*the sergeant at arms had told the BBC that on a request from you he would be prepared to allow cameras.*"[34] The same memo went on to recommend, however, that Admiralty House, the Prime Minister's temporary residence, would be best suited as the venue because it was more suitable to accommodate the media.

Immediately after that day's Prime Minister's Question Time the Prime Minister headed back to the Admiralty. The short meeting with Gagarin at 3.45 p.m. took place in a drawing room on the first floor of Admiralty House overlooking Horse Guards' Parade and with a clear view of St Paul's Cathedral. The Prime

6. A Delightful Fellow

Minister was accompanied by Heath Mason, the head of the Northern department of the Foreign Office, along with members of his personal staff. With Gagarin were General Kamanin, Boris Belitzky, Alexander Soldatov and Nikolay Denisov. The meeting with the Prime Minister received widespread coverage. Photographers, cameramen and journalists mingled outside the Admiralty. The Prime Minister spoke of *"the very great pleasure that the British people had taken in his visit"*. Gagarin said, *"it had been an overwhelming reception. People could not have been more friendly and cheerful"*.[35] With the public looking on, they waved from the first-floor window.

On behalf of Her Majesty's Government, the Prime Minister presented Gagarin with a 12-inch square silver salver of British workmanship (which was later inscribed with a message and sent to Moscow). In return, Gagarin presented the Prime Minister with a signed copy of his book *Road to the Stars* that had been published in Moscow on the day before. Speaking to reporters the Prime Minister later said about Gagarin, "*He's a delightful fellow*".[36] There was never going to be anything more than small talk at this occasion; the significance was in the photos on the front pages of newspapers around the world on the following day: pictures of Gagarin in his uniform, a member of the Soviet Armed Forces being welcomed in the heart of the British government. This was the political triumph that Khrushchev had calculated, sought and got.

Gagarin, still only twenty-seven and inexperienced in politics, had conducted himself with tact, courtesy and confidence. The guiding hand of his boss Kamanin was always nearby, but Gagarin did not need his assistance. The training, guidance and preparation in Moscow and at the Soviet Embassy in Kensington had been sufficient. Gagarin had proved his technical aptitude and courage as a fighter pilot prior to his selection as a cosmonaut candidate eighteen months earlier. But his engaging, personality, instinctive charm, and the natural and sincere smile that was constantly mentioned by eyewitnesses as his most

6. A Delightful Fellow

distinguishing trait were innate characteristics, not the result of training. The same characteristics had played a critical role in his selection as the first cosmonaut. As Patrick Moore[37] and Reg Turnill[38] had concluded, in picking Gagarin the Soviets had chosen the right man. Gagarin seemed to conduct himself effortlessly and appeared to enjoy it. At the end of the short visit, the Prime Minister came out to the front step and shook hands with Gagarin and other members of the Soviet Communist Party as the cameramen recorded the moment.

Kamanin was aware of the international political ramifications. He was closely observing Gagarin throughout the meeting with the British Prime Minister and could claim a high degree of satisfaction when it was all over and they waved farewell.

The next planned stop was the nearby Department of Defence but at the time known as the Air Ministry. Whilst travelling along Whitehall, Gagarin made an unscheduled stop whilst passing the cenotaph – unscheduled, but a stop for which he had prepared. He got out of the car, walked over with a large wreath of scarlet carnations and roses, and placed it at the foot of the cenotaph. Stepping back, he saluted.[39] It was a fitting commemoration by a Soviet soldier on his way to the British Air Ministry of the dead of two World Wars in which Russian and British soldiers had fought a common enemy.

It was now a little after 4.00 p.m. and it was at the Air Ministry where one of the most remarkable meetings throughout Gagarin's five-day visit took place. During this meeting, perhaps because of the absence of cameras and journalists, Gagarin described the details of his flight, including the all-important detail of his ejection from his spacecraft and landing separately from it. He continued reluctantly, but repeatedly to lie about when asked in public.

The Air Ministry was founded soon after the first use of aircraft in warfare during the First World War. The formation of the Air Ministry was a recommendation in a report published on 17th August 1917 by General Smuts during the final stages of the war. On 2nd January 1918, the Air Ministry came into being with Lord Rothmere as the minister. This new ministry would have an equal standing with the Army and Navy in all matters of defence of the nation. In 1964 the Admiralty, which had the authority over the Navy, and the War Office, a department of government which had the responsibility for the British Army since the seventeenth century, merged into the Department of Defence.

Test Pilot to Test Pilot

In 1961, it was under the auspices of the Admiralty that one of the most noteworthy meetings of Gagarin's entire visit took place. Immediately after the meeting with the Prime Minister, Gagarin the pilot had a private meeting with another pilot. The Deputy Director of Naval Warfare, Captain Eric Brown, had been the chief test pilot at Farnborough. Although Brown is not certain, he believes the Admiralty had requested a meeting where a test-pilot to test-pilot discussion could take place. Brown considers Gagarin as a test pilot, but his historic spaceflight notwithstanding, compared with Brown's record, Gagarin had very little testing experience. Brown's achievements as a test pilot were well established and it is likely that it was with the knowledge of Brown's accomplishments that the Soviets agreed to such a meeting in the first place. Brown still holds the world record in deck landings (240) and the number of aircraft types flown (487). He also has several firsts (first deck landing of a twin engine aircraft, first deck landing of a jet engine aircraft), of which the Soviets and Gagarin would have been aware.

Some within the aviation industry consider Brown as the greatest ever test pilot.[40] As aircraft development approached the supersonic arena immediately after the Second World War,

6. A Delightful Fellow

Brown had tested aircraft up to Mach 0.9 and the sound barrier seemed to be in reach. He had been testing a secret high-performance aircraft designated as the Miles M52, a turbojet-powered supersonic research aircraft in development since 1943 that was suddenly dropped in 1946. Had it not been cancelled, it is very likely that Brown would have added first supersonic flight to his collection of firsts. In the event, Chuck Yeager in the USA claimed that achievement in 1947.

Figure 23 Captain Eric Melrose "Winkle" Brown (Courtesy Captain Brown)

Brown had been encouraged to take up aviation by his own father who qualified as a pilot during the First World War and also by a First World War Ace German pilot, Ernst Udet, whom he met whilst visiting Germany with his father. Udet took Brown for a flight and encouraged him to take up flying. Persuaded by Udet, Brown learned to fly whilst studying German in Edinburgh. Brown made several journeys to Germany and was in Germany on the day war was declared. The German language and pre-war German connections became useful during his career as a test pilot throughout the war.

In the immediate aftermath of the war, Brown was given the

6. A Delightful Fellow

responsibility *"to investigate certain valuable items of German research before the Russians could get to them or the Allied soldiers destroyed them by mistake"*.[41] During this period he interrogated key German aviation personnel, including Ernst Heinkel, Willy Messerschmitt, Hanna Reitsch and even Hermann Goering and Wernher von Braun, and tested fifty-five different German aircraft, some of which he helped to recover from Germany and surrounding countries. He later wrote about these aircraft, including technical details and cutaway diagrams, in several books. One called *Wings of the Luftwaffe* could serve as an ideal instructional book for trainee pilots. Many were later translated into Russian, but not at the time when Gagarin was undergoing his flight training in Orenburg.

Brown frequently hosted visits from the Russian air attaché, now retired, General Makhov, to accommodate requests to participate in demonstrations of new British aircraft or weapons. In 1963, in the role of assistant chief judge, he attended the annual Helicopter World championships in Viciebsk in Belarus, then one of the member states in the Soviet Union. Despite his own unique achievement, it is likely that Gagarin was aware of Brown's accomplishment as a test pilot and wanted to meet Brown as much as Brown wanted to meet him.

It was agreed that prior to the familiar reception with drinks and finger food, a meeting between the two pilots would take place in Brown's office within the Admiralty. Brown did not speak Russian, so Boris Belitzky, who had been at his side throughout his visit, accompanied Gagarin. Brown was an experienced communicator and understood the subtleties of acquiring information in sensitive situations where the protagonists may not want or be able to talk openly. Brown did not know Belitzky and had suspicions that Gagarin's interpreter would not execute his role sincerely. To allay his doubts that *"we would really get back what Gagarin said in his replies"*,[42] Brown invited a colleague who was a Russian-speaking naval interpreter but concealed his ability to interpret, and through a prearranged

6. A Delightful Fellow

subtle visual signal would indicate to Brown a confirmation that the Belitsky was translating accurately or not. By the end of the conversation, Brown and his naval interpreter had concluded that Boris Belitzky had interpreted accurately as far as they could tell.

Between the cenotaph and meeting at the Admiralty, Gagarin had changed into civilian clothes and arrived at Brown's office with his interpreter Belitzky. His assistant with a remarkable similar name of Captain E.G. Brown and their Russian-speaking naval interpreter accompanied Brown. The five men conducted the meeting in what turned out to be a frank and open atmosphere. Brown recalls that his first impression of Gagarin was that he appeared as "*a very simple citizen, certainly an unsophisticated, stockily built man and could have been a Russian farmer, but altogether I found him a delightful person*".[43] The private meeting would last about twenty minutes, so Brown had calculated that he had sufficient time for about nine or ten questions including the translations. He prepared the questions prior to the meeting and still has a record of the questions and answers. Using the written record from the time, Brown recalls the following conversation between himself and Gagarin as translated by Belitzky in the company of Brown's two colleagues:[44]

> **Question:** "What was your main impression during the flight?"
> **Answer:** "It was the beauty of the universe and in particular "our world" as I saw it from space."
> **Question:** "What was your main sensation during the flight?"
> **Answer:** "Initially the tremendous noise and acceleration on lift off but then, as I approached orbit it suddenly changed to serene peace and quiet in orbit. There was little sensation of speed. I felt I was sitting there drifting around. Only the rotating features of the hanging Earth below indicated the actual speed."

6. A Delightful Fellow

Question: "Were you concerned for your safety?"

Answer: "No, because a test pilot knows how to cope with fear. But I must confess I felt a little trepidation about the re-entry and the landing. I wasn't frightened but I was concerned. I knew that if anything would go wrong it would go wrong at that stage."

Question: "Had you much to do on the flight?"

Answer: "No. Not compared to a standard aircraft test pilot on Earth. I was a passenger for much of the time."

Question: "Would you sooner be an astronaut or an aircraft test pilot?"

Answer: "I'd by far sooner be an aircraft test pilot. For then I would have greater control over my fate." [*Brown did not expect to get a sincere response but considers that he did*]

Question: "Would you like to return to test flying?"

Answer: "I sincerely hope that I am allowed to do so."

Question: "During the flight, what sight on our planet impressed you the most? Was it the Great Wall of China or the Grand Canyon?"

Answer: "No. It was neither of them. What impressed me the most was the huge size of Mother Russia." [*Given that national boundaries are not apparent when viewed from space, Brown considers that this comment was probably scripted for his domestic audience*]

Question: "Did you have any problems with the re-entry, the bailout or landing procedure?"

Answer: "No, I did not; but although the heat of re-entry did not frighten me I was very conscious of it. I was very thankful that the surface wind was so light." [*They had to come down so far and then bail out of the spacecraft and then land by parachute. I think frankly he had a slight feeling that this was a rather basic way of doing it, whereas the Americans would land in water. I had a feeling that as test pilots we had quite a rapport the two of us and he actually looked as if he was enjoying it and not po-faced and thinking, "Oh God, I have to through all*

6. A Delightful Fellow

this"].
Question: Finally, I asked him, "what was your main feeling in having achieved such a historic flight in space?"
Answer: [*He thought about a bit about this and then said*] "I feel proud to show the world that Russia is leading the race in space technology."

Brown's record of that conversation is the only instance of Gagarin's unambiguous acknowledgement that he had ejected, as this is the only possible interpretation of Gagarin's remarks on the surface wind. When asked why Brown had never shared this recollection before he replied, "*no one has ever asked*".[45] Brown, a Royal Navy Pilot, was seconded as an Exchange Officer to the United States Naval Test Pilot School at Patuxent, Maryland during the 1950s, where he had befriended Alan Shepard who on 5th May 1961 became the second person in space. Despite this close connection to the space programme, Brown was unaware of the controversy surrounding Gagarin's landing and so did not appreciate the significance of Gagarin's response about his landing.

Following Gagarin's return, the Soviets submitted an application for the world altitude record. The requirements, set by the International Aeronautical Federation specified that the pilot land inside the craft in which he had taken off. Gagarin had ejected at 7 km altitude so he was instructed to lie, which he did except during the meeting with Brown. Perhaps he felt it was safe to speak candidly; after all, it was a very small private meeting of professional pilots and journalists, and cameras were completely absent.

Whilst Gagarin was in Britain, his colleague Ivan Borisenko received a grilling from the IAF officials in Paris, seeking evidence for the world record. Despite the absence of firm evidence in support of landing inside his spacecraft, the IAF officials gave up and awarded the altitude record to Gagarin.[46] Two years later in September 1963, Kamanin and Gagarin

6. A Delightful Fellow

visited the Eifel Tower whilst attending the 14th IAF congress to collect the 20,000 franc prize.[47]

At the end of the twenty minutes, the private meeting ended and Gagarin was ushered into a larger room to meet other naval officers and senior civil servants. Brown recalls that he detected an obvious sense of discomfort in Gagarin as he moved from a small private meeting to a larger public one hosted by the Secretary of the State for Air, Julian Amery. In addition to several senior civil servants, amongst those present were Air Chief Marshall Sir Thomas Pike, the Chief of the Air Staff, and Major General I.P. Efimov, the Soviet Military Attaché. As the reception ended, the Secretary of State presented the now non-smoking spaceman with a silver cigarette box. Unlike the gift of the Silver Salver from the Prime Minister, the cigarette box had already been engraved. It had the motto of the Royal Air Force, *Per ardua ad astra* ("Through Adversity to the Stars" or "Through Struggle to the Stars"). The Major said, "*he would treasure the box, which would remind him of the days when the RAF fought so gallantly alongside the Red Air Force*".[48] In exchange, Gagarin presented another signed copy of his recently published book.

In 1961, there were several Soviet affiliated organisations seeking time with Gagarin to help promote their communist ideals in the UK. The GB-USSR Association was a HMG sponsored organisation and had invited Gagarin to its event in Hyde Park Hotel on Thursday evening. For HMG, it was imperative that Gagarin attend this event rather than those organised by the Soviet government sponsored communist organisations.

To that end, a confidential memo dated 11th July 1961[49] encouraged the Prime Minister to meet the spaceman again, for the second time in the same day, to bolster support for the GB-USSR chairman, Sir Fitzroy MacLean. If senior government officials were to attend, then Gagarin was unlikely to snub the

6. A Delightful Fellow

invitation. Fearing that the Soviet Embassy would persuade Gagarin not to attend, thus undermining the GB-USSR Association, one such memo concluded that:

"We believe that we could ensure Gagarin attends by letting it be known to the Soviet Embassy that Ministers as well as distinguished scientists and engineers will be attending. The Foreign Secretary himself proposes to go and if the Prime Minister himself went too, the standing of the Association would be enhanced."[50]

Figure 24 Yuri Gagarin with the Prime Minister for the second time on 13th July 1961
(Courtesy RIA Novosti)

The GB-USSR reception held in Hyde Park Hotel started at 5:30 p.m. There was a surprisingly strong attendance of around 500 guests. On the face of it, it was another series of superficial handshakes, pictures and formal introductions. But deep down this was a clear success for the British diplomatic efforts; the profile of the GB-USSR Association was enhanced by the large

6. A Delightful Fellow

number of high profile attendees. The objective had been to ensure Gagarin came, and they had succeeded. The Prime Minister, convinced by the memo, made the time and met the cosmonaut twice in the same day.

This event had been at the end of the long hot and hectic third day of his visit. Perhaps he wanted a break from the constant chain of receptions, speeches and handshakes of the last eight hours, or maybe he or his handlers wanted to limit the political gain of the GB-USSR Association, an organisation of which his embassy and government did not approve: Gagarin was scheduled to attend another reception at the nearby Dorchester Hotel organised by the Muscovites-Association, a business organisation, but sent his apologies. The event in Hyde Park Hotel was due to end at 8:00 p.m. but the guest of honour left for the embassy at 7:30 p.m. with a view to resting. In the event, he changed into civilian clothes and squeezed in a ninety-minute night-time sight tour of London instead.[51]

7. Communism and Royalty

The last-minute invitation to Buckingham Palace required that Gagarin add another day to his planned visit to Britain. This resulted in unexpected additional time that gave him an opportunity to make a pilgrimage that so many others from the Soviet Union had made. A 12-km drive north from Buckingham Palace, cutting through the centre of the Hampstead Heath, is Highgate Cemetery, full of popular attractions, including Karl Marx's tomb. On top of a 2 metre high pillar is a bust of Marx by the Liverpool-born sculptor Laurence Bradshaw. The tomb bears two inscriptions, "Workers from all lands unite", the final line from the *Communist Manifesto*, and "The philosophers have only interpreted the world in various ways – the point however is to change it", from Marx's collection of other work published only after his death. What Gagarin saw was not the minor original but an imposing tomb that had been rebuilt in 1954 by the British Communist Party.

Karl Heinrich Marx contributed to several fields of study, including philosophy and sociology. He was born on 5th May 1883 in Trier, Germany's oldest city, on the banks of the Mosel River close to the border with Luxemburg. He wrote many books but he is remembered primarily for two, *The Communist Manifesto* and *Das Capital* that he wrote with his fellow German Friedrich Engels.

Despite his own middle-class origins, in his writings he is critical of the system of society where a privileged minority in society own the majority of the wealth at the expense of the majority. This system of capitalism, he concluded, was run by the middle and upper classes to the detriment of the working class and he predicted that its inherent excesses would lead to self-destruction. A society governed by the working class would follow, and eventually a system of classless state government he called Communism would emerge.

7. Communism and Royalty

During his lifetime Marx did not have the international recognition he has today. His writings became the bedrock of enduring political systems of many countries in the twentieth century, and some survive into the twenty-first, most notably Cuba and North Korea. China's politics of the twenty-first century has incorporated so many aspects of capitalism that it is no longer a pure Communist state.

Marx was not the father of the Soviet Union as Mahatma Gandhi is considered to be of India and Mustafa Kemal Ataturk of Turkey. Marx was not even Russian. Over time his name has inextricably been linked with the Soviet Union because it was founded on the principles of socialism that Marx first articulated. In 1917, the Russian Revolution brought the Russian Empire to an end, and following a civil war, the Soviet Union was founded in 1922, a union which lasted until 1991. At its height, the Soviet Union was a union of fifteen republics, but it extended its influence on the larger Warsaw Pact countries along its western border. As Hungary discovered in 1956, the union would enforce its policies with brutal military might against those that challenged it.

The Soviet Union was a one-party state led immediately after its inception by Vladimir Ilyich Lenin until his death in 1924. He founded the USSR as the first socialist state, based on the theories of Karl Marx, but his ill health, resulting from a series of strokes, limited what he could have achieved. Having survived a revolution, civil war and assassination attempts, he died from illness when the Union was only two years old. Until its demise in 1991, the people of the Soviet Union had venerated both Marx and Lenin.

For Gagarin, as with others in the Soviet Union, Marx and Lenin were not vague distant characters in history but individuals who had a tangible impact on their everyday life. The pervasive indoctrination inherent in the Soviet propaganda is apparent in Gagarin's so-called autobiography *Road to the Stars*. It was an

7. Communism and Royalty

integral part of the Soviet education system, as Gagarin says, *"my class mates drew portraits of Lenin and wrote verses about him"*.[1] He also describes the day he took the compulsory oath of allegiance as a young soldier: *"January 8th 1956 is a day I shall remember as long as I live... Lifting my head I saw [a Lenin portrait on the wall] the penetrating eyes of Lenin who seemed to be listening to us young soldiers taking the oath."*[2] The commitment to socialism is reinforced at every opportunity: describing his oath of allegiance he says, it *"is the powerful meaningful expression of the Soviet man's love for his socialist country"*.

The spectators, including press and private cameramen, housewives, local workers and school children had congregated at Highgate Cemetery before the cosmonaut arrived. Gagarin and his party had parked in the cemetery car park from where the tomb is a short walk. Many in the crowd came from the opposite direction of Swain's Lane along the curved path to the tomb.

When Gagarin arrived at the cemetery, it was windy, the rain had almost stopped, but it was still overcast with the occasional drizzle. In contrast to his appearances elsewhere, the crowd here was smaller and quieter and required only a handful of policemen to keep in order. Gagarin walked down the path along the eastern side of the cemetery to participate in what at one level was an odd vision, a Major from the Soviet armed forces saluting a German scholar and revolutionary so soon following a war between their two nations giving rise to the largest loss of life in the twentieth century.

The tomb already had a couple of large wreaths from recent visits, the flowers still fresh. They had been repositioned before Gagarin arrived to make room for another one directly under the bust on the tall pedestal. Gagarin approached with an excessively large, almost clumsily so, wreath of red roses and white carnations and included a card with a message in Russian.

He delicately placed the wreath vertically against the tomb before taking a few steps back. Without a pause he saluted an icon of his country. Standing alongside, his Soviet colleagues including the ambassador and General Kamanin paused expressionless to take in the significance of being so close to the last resting place of someone they had only read about in books. This was the third wreath in three days, where the absence of Gagarin's extraordinary infectious smile captured the poignancy of the ceremony.

A Slice of Good Luck

Amongst the students watching Gagarin's salute were two eleven year old school boys, Philip Butler and John Zarnecki from Highgate School a kilometre away. They arrived at the cemetery via Swain's Lane and then onto the path leading to Marx's tomb. Their school motto was *Altiora In Votis* (Latin: *I pray for the higher things*) and on that day, in part triggered by this encounter with someone who had truly accomplished a feat of a superhero, each in his own way would be inspired to go on to perform higher things.

Philip Buckler, now the Very Reverend Philip Buckler, Dean of Lincoln Cathedral, remembers that he was comforted on arrival at the cemetery by the sight of a small crowd with a modest police presence, and he knew he was not too late. After only a short wait he saw a group of people coming down the path with the diminutive figure in a military uniform clearly standing out. The Dean remembers unmistakably the very moment Gagarin passed by him on his way to the tomb:

"I felt huge excitement at seeing someone so much in the news (what today we should call a 'celebrity' – but how very paltry today's celebrities seem when set alongside the achievements of a person like Gagarin). However, it was his return journey up the path that is the more vivid in my mind. It seemed almost

7. Communism and Royalty

unbelievable that we were within touching distance of this exotic figure, someone whose bravery had become legendary, who had dared to go into space, who had been and done something so beyond our comprehension. One wondered just what he had seen, what mysteries had he explored?"[3]

It is rare to be able to point to a specific chance meeting as a life changing experience. However, it did play its part. The memory of that brief visit has stayed with the Dean as a moment of profound significance perhaps shaping some of his qualities of open-mindedness, humility and sensibility.

As the Dean explored through his studies in theology the extent and limitations of human understanding, John Zarnecki took the road of science. He recalled this encounter in an interview *"I was gob smacked. If there was a day that changed my life that was as close as it came!"*[4] In his 2007 Open University Lecture entitled "Half a Century in Space" to commemorate the 50th anniversary of Sputnik, he observed:

"Now, my first slice of good luck, I think, happened just to be growing up in the 1960s. It wasn't just the Beatles, the Rolling Stones, Pink Floyd, England winning the World Cup – I saw all of those live. But I really became hooked on space exploration in 1961 and a chance encounter with this man – Yuri Gagarin – the first man in space and overnight the most famous man on Earth".

Zarnecki has not made it to space himself but some hardware his team designed and built did. One of the numerous projects that he has been involved with was a collection of instruments that landed on the surface of Saturn's largest moon, Titan. It took seven years to build and test the instruments, and after another seven for the journey to its destination, it arrived on Titan on January 14th 2005. After a two-and-a-half-hour descent on parachute through Titan's thick atmosphere the probe landed on a dried-up lake bed and continued to transmit data for an

7. Communism and Royalty

additional seventy-two minutes. The probe sent images of lakes, mountains and valleys of an alien landscape that looks remarkably like that of Earth.

It is impossible to say if the experience of seeing the world's first spaceman was a transformational experience that irrevocably set John Zarnecki and Philip Buckler on the course of their respective professions. As Gagarin illustrated through his short but spectacularly eventful life, we are all the products of our experiences. For a few days during the hot summer of 1961, many had the opportunity to be in the presence of the only person in human history to that day that had orbited the Earth.

Confidential memos, now declassified, between the Home Office and Buckingham Palace reveal the last-minute preparation of the spaceman and his companions' visit to Buckingham Palace. The Home Secretary requested that in addition to Gagarin, the ambassador and Gagarin's interpreter, another two be also added to the list for the Friday lunch party, General Kamanin and Gagarin's Boswell, Denisov. The announcement of Gagarin's invitation to meet the monarch had attracted widespread publicity, but it was an invitation to a lunch that had already been planned and not something designed especially for him. Despite the appearances in the press this was not an official visit, and the Palace went out of its way to stress that and the diversity of guests invited. The Palace informed the Home Office that it should:

"explain to the ambassador about the composition of the Luncheon party and to make it clear that, far from being an official one, it is to be made up of a wide variety of her Majesty's subjects. I think that the Russian guests should understand that before they arrive; otherwise they may well be perplexed. I must also point out that it is impossible to add further to the numbers invited without great domestic inconvenience and reorganising the party on a different scale."[5]

7. Communism and Royalty

In response, the Foreign Office memo confirmed that an additional non-eating interpreter, John Morgan from the Far Eastern Department, would attend as interpreter for the Queen, in addition to Gagarin's own interpreter, Belitzky. The Queen was provided with "personality reports" on all the visitors with the exception of Gagarin and the Soviet Ambassador; an arrival time at the Palace was established for 12.50 p.m.[6]

Crowds had lined the route to Buckingham Palace prior to his arrival. It was a dry and warm Friday afternoon as the two cars escorted by eight police motorbikes and mounted police approached the palace from the Mall, curving along Spur Road into the centre gate. The crowd make up was similar to that Gagarin had experienced at the Tower of London. Several thousand had lined up on both sides of Spur Road and around the Victoria Memorial at the centre of Spur Road. Many, especially children, ran along the car as it slowed. The roofs of both cars were down, so only those close by actually managed to see the cosmonaut. Crowd control outside the palace was something with which the police were familiar. Besides, it was not an unruly crowd; the police had only to escort him through to the centre gate into the forecourt of the Palace. Gagarin was not going to stop. As he drove up to the gates, several photographers took pictures with flash despite the daylight.

Buckingham Palace, originally known as Buckingham House (and still unofficially referred to by some as Buck House), has been the official home of the British monarch since the accession of Queen Victoria in 1837. It has almost eight hundred rooms with one hundred and eighty-eight bedrooms, ninety-two offices, and seventy-eight bathrooms just for the staff who maintain the large complex laid out over almost thirteen thousand square metres. The palace garden, which includes a lake, is the largest in central London. The palace has a ballroom and a swimming pool, a throne room, several drawing rooms, a picture gallery, a music room and the state dining room.

7. Communism and Royalty

Below the state dining room, there are a number of semi-state rooms in the west wing used for formal but smaller events. The invitations for the majority of the guests had gone out long before Gagarin was invited, so the additional five visitors were a genuine administrative concern for the palace staff that had to arrange the seating plan. Possibly in deference to the Russian guests, Room 1844 with its unique Russian connection had been selected for the lunch event. During their visit to London in 1844 Queen Victoria hosted the Emperor Nicholas I and his Empress in that room – potentially a royal faux pas given that the Russian guests represented the Soviet Union which through a bloody revolution and civil war had overthrown the Empire. Perhaps the guests did not feel they were in a position to express any dissatisfaction if they felt it.

As this was not a state occasion, the diverse guest list included Joanna Kelly, governor of Holloway prison, the architect Sir William Holoford, the mountaineer Sir John Hunt and the comedian Bud Flanagan. Gagarin sat as guest of honour on the Queen's right-hand side, and each with their respective interpreters on either side. The Duke of Edinburgh sat opposite the Queen with the Soviet ambassador on his right.

Gagarin had lived with his parents and then in dormitories during his training at the technical college and in Orenburg while undergoing his military pilot's training. It was only in the autumn of 1957 when he qualified as an air force lieutenant, married and was posted to the Arctic Circle that he had a modest apartment he could finally call his own. This palatial setting would have been a sharp contrast. At the outset of his tour, perhaps, he would have been taken aback by the elegant architecture, expensive paintings and sheer sumptuousness of a palace but he did not show it. On the menu for lunch was poached egg, cheese sauce and saddle of lamb with all the trimmings. Bud Flanagan commented that Gagarin *"gave the food what for. My god, he must have been hungry, that boy!"*[7]

7. Communism and Royalty

Later he commented, *"Yuri, he is really nice bloke. Very unassuming"*.

In a book due to be published in July 2011, *The Life of Remarkable People* by Lev Danilkin, Vladimir Lebedev recalls that following his return to Moscow Gagarin had told him of his confusion at the dining table. Seeing the large array of cutlery, he turned for help to the Queen who replied, *"My dear Mr Gagarin, I was born and brought up in this palace, but believe me, I still don't know in which order I should use all these forks and knives."* Lebedev also claims that Gagarin went further:

"he wanted so much to be sure it was a real queen that he touched her under the table, slightly above the knee." In response to being touched, *"the Queen just smiled and carried on drinking her coffee"*.[8]

Gagarin loved practical jokes. His niece Tamara Filatova remembers that Gagarin liked pranks and was always able to get people to forgive him. *"No-one was able to resist his smile! And girls always liked him, too."*[9] Top secret papers declassified to mark the 50th anniversary of his flight illustrate Gagarin's humour on the morning of his historic space flight. During the final preparations before launch, the chief designer reiterates the food rations on board.

"You've got sausage, candy and jam to go with the tea," Korolev went on. *"Sixty-three pieces – you'll get fat! When you get back today, eat everything right away."* Gagarin joked back: *"The main thing is that there is sausage – to go with the moonshine."*[10]

It is possible that Lebedev's account could be true, but unlikely. Gagarin was the guest of the monarch of a nation with very long traditions, as the opulence of his environment reminded him. More significantly, he was in the company of high-ranking

7. Communism and Royalty

senior Soviet personnel that he held in very high regard. It is unlikely he would have risked offending a distinguished host on his first official meeting.

The schedule for what was left of the rest of the day was a monotonous one involving repeated trips to the Embassy and Earl's Court. The first visit to Earl's Court was to speak to school children. They had been part of the crowds everywhere he had gone, but this was the first opportunity for him to address them directly, once again via his translator Boris Belitzky. He would return one more time to Earl's Court for a live BBC TV interview before returning finally to the Embassy well after 10.00 p.m.

The students dressed in their school uniform jackets and ties were asked to welcome the spaceman with a "Wembley cheer" (a loud cheer just as one that accompanies a goal scored at Wembley football stadium), and then as their space hero began to speak, they hushed into a deep silence that their teachers must have envied. Gagarin began by thanking them "*I am very happy to have the opportunity to meet so many of you whilst here in London, in Manchester and to see you in the streets to wave to you and to hear from you.*"[11] He also thanked many that had written to him and expressed his hope that one day, like him, they too would experience spaceflight if not as cosmonauts then as passengers.

Remembering the decisive role of his school teacher Lev Mikhailovich Bespalov, a former pilot, during his early schooling Gagarin reminded the eager students in front of him of the importance of studying. He concluded, "*Space flying requires a certain amount of sound knowledge in sciences, mathematics and other subjects. I wish you well in your studies*".[12]

7. Communism and Royalty

Danger of Push Button War

The British-Soviet Friendship Society (BSFS) was one of the several pro-Soviet groups with the objective to foster better east-west relations during the period of the Cold War. A consequence of its political credentials resulted in the BSFS promoted by the Soviets but scorned as a "communist front" organisation by the HMG. Gagarin made an appearance at the BSFS affiliated organisation the SCRSS based in Kensington Square, a short distance from the embassy. Members of both SCRSS and BSFS were present during a garden reception on a warm summer evening, the final one of his UK visit. Gagarin motivated by his childhood experiences of German occupation, made, according to the Daily Worker, "*what was probably his most important speech*".[13] The Major had said that "*If people everywhere worked for peace as the society and its equivalent in Russia were doing then the danger of push button war would not arise*".

The press coverage of the meeting with the Queen was a predicament for her Government. It naturally attracted particularly heavy media coverage with headlines about the warm, rapturous and enthusiastic welcome the cosmonaut received from the British people. A Foreign Office memo expresses the concern that the British Government was left "*facing a renewed wave of accusations that we are soft towards Moscow, soft on Berlin and soft all round*".[14] The same memo includes a reference to the comedian Bud Flanagan whose apolitical comments diluted the political tensions of which he was probably unaware.

John Platts-Mills, a former Labour MP and Rhodes Scholar originally from New Zealand, was the chairman of the BSFS and presented the cosmonaut with a fibreglass travelling case. He acknowledged the boost that "*his achievements and personality had brought Soviet British friendship to a new high*".

7. Communism and Royalty

In the midst of the massive turnout and the universal adulation since his arrival, which surprised the hosts as much as the visitors, perhaps Gagarin felt he really had made an impact. After several days of repeating the mantra of peace, friendship and cooperation, if he had not achieved a tangible political shift, perhaps he had succeeded in changing the unfounded Western perception of the hostile aggressive Soviet people to something that reflected reality more accurately.

For the final formal engagement of his last full day in Britain Gagarin went once more to Earl's Court, where the Soviet Trade Fair was still running and open until 10.00 p.m. In a simple studio with two chairs and a desk at one end for Gagarin and his interpreter Belitzky, with the three questioners, Richard Dimbleby, Tom Margerison (science editor of the *Sunday Times*) and Yuri Fokin of the Soviet Television Service, on the other. Gagarin had a pen and paper, but appeared to doodle between questions rather than take any serious notes.

Gagarin's familiar smile appeared to obscure his underlying anxiety of a live interview on British TV. When not smiling Gagarin looked nervous and anxious: this would be his first extended live interview on British television. Unlike the superficial questions he fielded during his first press conference, the members of this panel were experienced professionals, and given that this was live, there was no second chance.

All of Gagarin's questions and answers are translated by Boris Belitzky. The edited account of that interview below starts with the first question about "fear" at take-off, but Dimbleby, the seasoned interviewer, did not use the word fear itself.[15]

> **Richard Dimbleby:** We have a saying here that when you are nervous about something you have butterflies in your stomach or butterflies in your tummy. Can you really honestly say that you did not have any butterflies in your

7. Communism and Royalty

tummy?

Yuri Gagarin: Yes, I can assure I had no butterflies or moths or anything in my tummy.

Tom Margerison: Your American astronaut colleague spent an unfortunate 3 or 4 hours in his spaceship before he started off. Did you have a comparable wait before your take off?

Yuri Gagarin: We were not in the same position. There was no need for me to spend several hours before the takeoff. The brief period of time I spent in the spaceship on the ground before take-off I spent in normal condition and I think the scientists can show that in the data recording at the time. There were no grounds for me to feel anxious at that time or later.

Richard Dimbleby: When are we going to see the colour film of the launch that everyone here is anxious to see?

Yuri Gagarin: It is difficult for me to give you an exact time. It does not depend on me but it is being shown in the Soviet Union now.

Tom Margerison: Could you give us some idea of what it's like being in the spaceship? How much room did you have in the spaceship?

Yuri Gagarin: Yes, it was quiet roomy. In fact it was much roomier than an aircraft cockpit.

Richard Dimbleby: One small mystery that you can help clear up. On the day that Moscow radio announced your flight, there appeared a report in our communist newspaper, *Daily Worker*, a report indicating that a flight had been completed successfully and the flyer had returned safely. That report from Moscow was dated a day before. This created the impression that another flight had taken place and you had flown second.

Yuri Gagarin: I can assure you quite authoritatively that the correspondent of that newspaper felt he was better informed than the actual people in charge of this work in the Soviet Union. No flight of this kind had taken place in the Soviet Union or in fact in any other country. The flight

7. Communism and Royalty

made on April 12th was the first flight in history of this kind, the first manned spaceflight ever.

The full twenty-six-minute interview was transmitted live at 9:30 p.m. on Friday 14th July 1961. Most of the population of Britain thus was able to get an insight into the man that history remembers as the first in space through this interview.

The airport was the final stage of his five-day visit and the final opportunity for anyone who had wanted to see him in person. Buses to London's Heathrow airport were full on the morning of Saturday 15th July. The Major was returning to Moscow on a scheduled flight, SU032 at 12:35 p.m. The twin-engine turbo powered medium range jet was one of the first jet airliners to go into regular commercial service, operated by Aeroflot, at the time the world's largest airline.

The day had started outside the front door of the Soviet Embassy in 3 Kensington Palace Gardens. It is surrounded by other embassies and high commissions with outer tennis courts and occasional helipads in a leafy suburb of west London. The entrance from the main road is through a narrow gate between two short pillars opening out to a very small courtyard. The main entrance to the building is on a tiny raised veranda with steps left and right leading down. The Embassy had been Gagarin's base since arrival. He had started and ended each of his five days in Britain from here. Now it was the final departure. He did not know it, but he would never return.

About 11:00 a.m. he came out of the front door into a congested compact forecourt with people sardined into the place where his open top Rolls Royce with the personalised number plate was parked along with the police motorbike outriders. The very short distance from the door down to the steps was crammed with well-wishers and photographers taking advantage of every raised platform. Slowly, assisted by the police including the helmeted motorbike escorts, Gagarin was guided into his car. He was

7. Communism and Royalty

accompanied once again by the Soviet ambassador, General Kamanin and his interpreter Belitsky in the front seat. Gagarin stood as his driver carefully moved the car forward to encourage the crowd to make room.

Initially at a snail's pace, the car gradually moved towards the gate leading out to the road, still with many people dangerously close by, waving, taking photographs and even trying to reach out for a last-minute handshake. The crowd had spilt out into the main road and a few had climbed the small boundary wall for a better view. It was a dry bright morning, and some in the crowd had their children on their shoulders for a rare opportunity to see a man whose achievements were beyond their comprehension.

The route to the airport was once again lined with crowds, only this time the waving hands carried a message of farewell. At the airport, the crowds mingled with passengers, their relatives and with the Saturday morning's crop of plane-spotters. This was the first non-working day of Gagarin's visit, which bolstered the numbers attending along the route and at the airport to bid the spaceman farewell. Gagarin had expressed his sincere appreciation of the welcome he had received from the British people during the previous night's live BBC TV interview. He now repeated it at a short improvised press conference at the airport, his closing event of his visit to Britain. This would be his interpreter Boris Belitzky's final event supporting Gagarin. In this press conference, Gagarin said "*I carry back with me the most warm memories of the friendly reception given me and of the hospitality in Great Britain.*"[16]

The police had put barriers to maintain a clear path along which Gagarin and his party could walk unimpeded towards the waiting aircraft. He walked towards the tarmac looking left and right, waving back at the large crowd looking on. Government and airport officials lined up at the bottom of the aircraft steps formally bid farewell to the army Major. Mr Francis Turnbull, secretary of the Office of the Minister for Science and Air

7. Communism and Royalty

Marshall Sir Ronald Lees, who had welcomed him on the previous Tuesday, were present to say farewell. Airport workers sneaked in more handshakes as he ascended the steps.

On leaving the Soviet Embassy on this final day of his five-day visit he wore on his left lapel his gold medal, the symbol of his honorary membership of the Amalgamated Union of Foundry Workers he had received during his visit to Manchester. Perhaps he was expressing the depth of the impact of his only meeting with working people. On arrival, back in Moscow he was asked what he liked most about his visit to Britain, to which he replied, "*The people*".[17]

8. A Smile that Changed the World?

The widely held belief that the space age began with the launch of Sputnik on 4th October 1957 is not one shared by Neil Armstrong, the first person to set foot on the Moon. He asserted in 2004[1] that it began with a research suggestion in 1950, which came to be known as the International Geophysical Year (IGY). Every eleven years the Sun goes through a cycle of increased solar radiation that has direct climatic impact here on Earth. A group of American scientists proposed in 1950 an international programme to study the Earth's upper atmosphere for the then next solar maxima around 1957/8. The research programme became known as the IGY and would actually run from 1st July 1957 to 31st December 1958. This was the first instance of a proposal in the USA that would hasten the work of rocket development, in this case Korolev's missiles, in the USSR.[2]

During a face-to-face meeting in 1956, Korolev reminded Khrushchev that:

"the United States had stepped up its satellite programme, but that compared to the "skinny" U.S. launch vehicle, the Soviet R-7 could significantly outdo that project in terms of the mass of the satellite. In closing, he added that the costs for such a project would be meagre, because the basic expense for the launcher was already allocated in the R-7 booster."[3]

Following the American developments and use of nuclear weapons in 1945, the Soviets successfully completed their first nuclear test in 1949 and realised that by using missile technology as a delivery vehicle, the requirement of big expensive bombers would become less significant. Sergei Pavlovich Korolev's design bureau had been tasked to design, build and test missiles for the Soviet military, and Korolev personally convinced Khrushchev that he could realise Tsiolkovsky's vision by building a small satellite and launching

8. A Smile that Changed the World?

it into orbit on his R7 without undermining the military programme. The Soviets' failure to design small lightweight nuclear bombs necessitated the huge lifting capacity of Korolev's R7 rocket, which he successfully tested as the world's first intercontinental ballistic missile in the summer of 1957.

In the pursuit of his personal ambition of spaceflight, Korolev exploited the fear of falling behind the USA to secure large public funds and political commitment at the highest level. Wernher von Braun eloquently replicated this approach in America. Like a high-stake poker game, each side invested increasing amounts of capital, not only financial and political, but also in terms of national reputation and prestige. This tactic, which began in the early 1950s and continued to evolve through the 1960s into a mutually propagating strategy, led first to Sputnik, then to Gagarin's spaceflight, and culminated with a manned landing on the Moon with incredible haste.

Boris Chertok, a colleague of Korolev's, expressed his amazement that the Soviet Union succeeded with manned spaceflight. He contended *"that if Gagarin's flight on 12 April 1961 had ended in failure, U.S. Astronaut Neil A. Armstrong would not have set foot on the Moon on 20 July 1969."*[4] Bernard Lovell also expressed his incredulity with NASA's ambitions for the Moon before the decade was out. Writing in 1968, before NASA had conducted any tests in space of the Apollo spacecraft that would take them to the Moon, he wrote *"one feels that the maintenance of this timetable would represent a miracle of management and technology"*.[5]

A decade and a half after a war in which the Soviet Union lost 30 million of its citizens, it was ready for its and mankind's next adventure. The goal, borne out of fear and perhaps a national inferiority complex, was to realise Tsiolkovsky's dream of manned spaceflight. The USSR was well-endowed with people who were intelligent, industrious and, as Andrew Jenks suggests, a little reckless. *"Sensible people would not have strapped*

8. A Smile that Changed the World?

Gagarin on top of a rocket – just as they would not have attempted to put nuclear missiles in Cuba, grow corn in Siberia, compete with the United States in consumer goods, condemn Stalin's crimes, underwrite revolution in Africa and Asia, and promise to reach communist utopia in 1980."[6]

They had been reckless and succeeded; first with heavy lift rockets, then Sputnik and now with Gagarin.

Mission Accomplished

The world's first successful manned spaceflight was such a sudden surprise that even the Soviets did not anticipate the magnitude of the international impact. Two days after the flight, on 14th April, Gagarin was welcomed to Moscow's Vnukovo Airport in his Ilyushin-18 accompanied by seven MiG jets flown by his air force colleagues in formation, two on each wing and three behind. Hundreds of thousands of people waved and cheered along streets, in the squares and from the windows of the route from the airport to Red Square to welcome the country's newest hero. The Communist Party extracted the maximum political benefit. Although they had not televised the launch of Vostok, they provided live TV coverage of these epic national celebrations throughout Europe. Richard Dimbleby narrated the BBC's coverage as Gagarin's plane taxied to a stop for the triumphant red-carpet meeting with Khrushchev at the airport prior to the drive to Red Square. The overwhelming enthusiasm and celebration by the massive crowds in and on the road to Red Square was not surprising. The international reception, including that in the UK, was a revelation.

Korolev's identity, along with the names of the other nineteen cosmonauts, was a state secret.[7] Consequently, in addition to the injustice of his imprisonment during Stalin's reign, he was overlooked again. Now on the day of this accomplishment for which he more than any other single individual was responsible,

8. A Smile that Changed the World?

he was neglected once more. As Gagarin's achievement was being celebrated under the eyes of the international media, Korolev without an official car was stuck at the very end of the motorcade from the airport to Red Square. According to one account of a conversation with Korolev's daughter, Korolev at the end of that historic day, still without an official car, had to go home on the Metro.[8]

Korolev had been overlooked despite being a very strong candidate for the Nobel Prize following the successful launch of Sputnik in 1957 (and again with Vostok in 1961). For the award to be made, Korolev would have had to be publicly named, but Khrushchev, steeped in the overwhelmingly secretive traditions of the Soviet state, would not allow it.[9]

Following the celebrations at Red Square the first international press conference was scheduled at the prestigious Moscow Academy of Sciences. The Soviet government had brought together a network of scientific institutions as the Academy of Sciences in 1925 as single premier scientific institution in the Soviet Union. Initially based in Leningrad, the headquarters had moved to Moscow in 1934.

On 14th April 1961, the hall was filled to capacity with Soviet workers, academics, scientists and foreign press. In the bright lights of several cameras, Gagarin, accompanied by several academicians and Soviet civil servants, walked onto a stage with a large brightly lit bust of Lenin behind him.

Good Humoured Evasion

The BBC did not have an aerospace correspondent based in the Soviet Union, so they sent Reg Turnill, who was the aerospace correspondent covering the American space programme. Turnill was in London covering stories about the closure of American bases in the UK when he heard about Gagarin's flight. Turnill

8. A Smile that Changed the World?

recalls, *"A western journalist based in Moscow reporting on Soviet technology was simply not permitted in those days."*[10]

Turnill's visa application was processed swiftly and he turned up at the Academy of Sciences for the press conference that he would later describe as *"good-humoured evasion"*, but remembers it as, *"a phony press conference from beginning to end. An entirely choreographed occasion designed to humiliate the west and humiliate people like me as representatives of the west"*. Another British journalist, Peter Fairley, expressed a similar sentiment describing Gagarin as, *"a master of the art of evading the awkward questions"*.[11]

For Turnill, the humiliation started on his arrival at the press conference. Initially his ticket was dismissed and he was not allowed to enter. There was an almost complete absence of provision for the international press. He had to search for a seat amongst the "rent-a-mob" crowd of workers who had been invited to this special occasion. He eventually found a seat in the midst of a crowd of *"large heavily built nurses"*. The conference allowed only for pre-submitted written questions. He posted his questions into a box and returned to his seat. In an interview in early 2011, he describes his experience of the press conference:

The press conference was preceded by an hour and half of speeches, which were partly translated (by Boris Belitzky). I was worried I would miss my slot on Moscow Radio from where I would file my report to the BBC. I was always impressed by Yuri Gagarin. He was a very smart young man, bright and self-assured and yet not vain or arrogant. [He was] a very rare human being. He stood up and started to answer questions. We got what journalists call the classic idiot treatment. Once the lengthy preamble finished the Q&A started.

Turning to his fifty-year-old type written notes, Turnill returned to the question and answer session at the press conference.

8. A Smile that Changed the World?

Question: "Did you land inside the spacecraft or eject and land on a parachute?" – *This was very important to understand the sophistication of the spacecraft. The Soviets wanted to claim the altitude record. Gagarin was pulled down by [Anatoli] Blagnarovov who whispered in his ear. Then Gagarin stood up, smiled and said –*

Answer: "Whichever way it was, you can see it was successful." – *This triggered a roar of laughter and this set the tone of the press conference.*

Question: "When were you told you would be the first cosmonaut?"
Answer: "I was told in good time."

Question: "How many cosmonauts are there?"
Answer: "In accordance with the plan to conquer cosmic space, pilot cosmonauts are being trained. I believe there are more than enough."

Question: "When will the next spaceflight take place?"
Answer: "Our scientists and cosmonauts will undertake the next flight when necessary."

Question: "I can go and watch an American space launch whenever I want to. When will I be allowed to watch a Soviet launch at Baikonur?"

There was a long answer in Russian and the translation was muffled by laughter and jeering from the crowd, but the gist of the answer was that they certainly weren't going to let the western spies into see and copy their advanced rockets.

At another time, I would have gone to town on it but it was not the right occasion. Everyone was in a congratulatory mood and it was a great achievement. One could not really knock it.

8. A Smile that Changed the World?

It was the most phony press conference I have ever attended. Although I put the best gloss on it at the time, it was a really unpleasant experience.

Other questions asked during this space conference included,[12] "*Will the photographs taken of Earth from space be published?*" Gagarin replied that no photographs were taken so none would be published. The only camera onboard was the one on the inside of the Vostok trained on him. On another question about the delay between him landing and the arrival of the recovery party, Gagarin replied that they arrived almost at the same time that he had landed. Another questioner asked if Gagarin's Vostok could be reused. Gagarin replied that the engineers were best placed to answer that question but he considered that his Vostok could be reused.

One of the questions addressed the issue of food. What had he eaten during the flight? "*Special food designed by the Academy of Medical Science*", he replied. Gagarin had eaten food by squeezing it out of tubes directly into his mouth. The flight was very short, he did not get hungry, but eating and drinking were part of his mission plan. It would be important to gather information on eating and drinking in the weightlessness. Missions lasting several days were already in the pipeline. His Vostok had food and water provision for up to ten days. This was in part Plan B, a contingency that if the designated rocket to return him to Earth failed, atmospheric drag over a period of ten days would cause the spacecraft to re-enter Earth automatically. Recent calculations suggest that Gagarin's actual orbit was higher than intended and thus it would have taken longer than the ten days to degenerate and allow him to re-enter automatically. Fortunately, this contingency was not required.

Unlike his American colleagues, Gagarin, considered too valuable an asset, was in effect banned from piloting an aircraft immediately after his flight in 1961. However, after years of

8. A Smile that Changed the World?

media interviews, ceremonial presentations and receptions, Gagarin gradually returned to spaceflight. In April 1967, the Soviets launched a new spacecraft on an ill-prepared mission piloted by Vladimir Komarov, who was killed when the parachutes failed to open properly after re-entry. During the extended period of review and redesign that followed, Gagarin and other cosmonauts focused on their training. Gagarin, still burdened with his media and ceremonial obligations, would frequently have long gaps in his flight training. Following a hiatus of three months, Gagarin had to regain flight readiness by a routine flight under the supervision of an instructor. It was scheduled for 27th March 1968.

In preparation for his supervised flight Gagarin attended a preliminary training session along with fellow cosmonaut Andrian Nikolayev on March 26th. After the training, as they walked together, Gagarin and Nikolayev met up with Nikolayev's wife Valentina and their daughter Alyona.[13] Valentina describes the fatherly attention Gagarin paid Alyona on that day:

Shortly before Yuri and Andrian came up we had a little accident. Alyona slipped and fell into a puddle and her white coat looked a sight. The girl began to cry. Yuri picked her up in his arm and asked, "Who's hurt such a nice little girl?" Alyona cried even louder. Yuri hugged her and said in a soft tender voice, "Don't cry Alyona, this is easy to mend."[14]

Valentina asked about Gagarin's wife (also called Valentina) who she discovered was still in hospital. Valentina said that they would go and visit her on the 28th.

On 27th March, with far from ideal weather conditions, Gagarin took off in the front seat of his dual seat MiG-15 with a very experienced and decorated instructor, Vladimir Seregin. His friend and colleague Alexi Leonov was in command of a parachute training session preparing cosmonauts for the Soviet

8. A Smile that Changed the World?

lunar mission about 80 km from the airfield that Gagarin had used for take-off. The weather worsened and Leonov cancelled the training. Whilst waiting for permission to return, Leonov recalls he heard two loud booms in the distance in quick succession, about two seconds apart. Over the radio he heard the air traffic controller repeatedly calling Gagarin's flight "241"[15] but he never got a response.[16] The MiG had crashed moments after Leonov had heard the two booms, killing both pilots instantly.

The investigation team discovered remnants of a hot air balloon near the crash site and concluded Gagarin's MiG had collided with the balloon, entered an unrecoverable dive and crashed. Other theories of a bird strike, mechanical or instrumentation failure, and even that Gagarin had been drinking prior to the flight, circulated at the time. Leonov was convinced that the two booms he had heard were that of a supersonic boom followed by a crash. Gagarin's MiG-15 was not capable of supersonic flight, but a new SU-15 also in the area that day was. Leonov believes that the SU-15, violating its flight level restriction of 10,000m, had descended to 4,200m, to within a few metres of Gagarin's MiG, which the poor weather had prevented the SU-15 pilot from seeing. The resulting turbulence put the MiG into a dive from which it did not recover.

Gagarin and Seregin could have ejected but neither did. The MiG-15 used two in-line seats with Gagarin in front of Seregin. The standard procedure for emergency ejection is for the instructor in the rear seat to eject first and then the pilot in the front.[17] Ejecting simultaneously has too high a risk of collision during ejection or of the two parachutes interfering with each other. Had Gagarin gone against the standard procedure and ejected first, the exhaust from his ejection seat would have shattered Seregin's canopy, undermining Seregin's chances of ejecting. Seregin, mindful of the national significance of his charge, could not have envisaged ejecting first, as procedure required, and abandoning Gagarin. Both men were friends as well as colleagues and had the highest mutual respect for each other. In the brief moments they could have ejected each put the life of the other first and both perished.

A recent, unverified, newspaper story, published on 8th January 2010 in the *Daily Telegraph* by a retired Soviet air force Colonel, Igor Kuznetsov, claimed to have solved the mystery. He concluded that Gagarin panicked after realising that an air vent had been open. In the ensuing dive, Gagarin lost consciousness resulting in the crash.

The Soviet obsession with secrecy[18] restricted the nature of the investigation and the sharing of its findings. Twenty-five years later, once the investigation documents were declassified, Leonov saw the evidence he had submitted at the time. The time interval between the two booms in his testimony of two seconds had been changed, by an obviously different handwriting, to twenty seconds. Leonov is convinced the crash was caused by the pilot of a supersonic SU-15 flying too close to Gagarin's MiG at a much lower altitude than he had clearance for.[19]

Cold War and Peace

In the time between the end of the Second World War and Gagarin's arrival in London, the USSR, a former ally, had become a Cold War enemy. This remarkable turnaround in the space of sixteen years had started even before the end of the Second Word War, as the allies scrambled to acquire for themselves the fruits of German aviation, rocket technology and research in the closing days of the war. The Americans through their "Operation Paperclip" sought out the remaining V2 rockets and the team that built them. They succeeded, primarily because the team leader Wernher von Braun wanted the Americans as much as they wanted him.

Following the war, a series of events over a decade and a half

8. A Smile that Changed the World?

laid the foundation for the tense East-West relations that prevailed when Gagarin arrived in the UK.

May 8th, Victory in Europe (May 9th in the USSR due to the different time zone) marked the end of the Second World War in mainland Europe.[20] During the last two weeks of July 1945, as the war against Japan neared its apocalyptic end with the first use of nuclear weapons, the long and complex Potsdam Conference concluded the future of Germany. It was to be completely disarmed, demilitarised and divided into the already agreed four zones of occupation controlled by the allies – UK, USA, France and USSR. This division persisted until 1949 when the three western sectors controlled by the USA, UK and France merged to form West Germany with Bonn as its capital, and the USSR zone became East Germany with East Berlin as its capital.

Disagreement between the allies over Germany's post-war future triggered the first crises of the Cold War. The Soviets blocked access to road and rail routes through their zone of occupation in East Germany to West Berlin. Throughout the period of the Blockade (24th June 1948 – 12th May 1949), all supplies to West Berlin were delivered by air.

This failure to demilitarise and disarm became a symptom and a cause for the Cold War. The unsuccessful East German revolt against Soviet occupation on 17th June 1953 ended any hope of a future agreement and rendered a united Germany another casualty of the Cold War. The subsequent development of military technology by USA and USSR of long-range missiles, high altitude spy planes and development of nuclear weapons further divided the former allies.

In the first few months of his presidency in 1961, Kennedy faced a series of major international crises. Communism and capitalism collided in post war Germany. The economics of a divided city (a capitalist West Berlin and communist East Berlin)

8. A Smile that Changed the World?

with two currencies was unsustainable. Some Berliners chose to study in the East (where it was free) and work in the West (where it was more rewarding). A month after Kennedy's inauguration, France became the fourth nuclear power conducting its first nuclear test on 13th February 1961. Agreements on nuclear disarmament and testing in discussion since the 1950s had become stuck, and nuclear stockpiles on each side grew. New in office, Kennedy's administration faced two new challenges. First, Gagarin's successful spaceflight on 12th April 1961, followed a few days later by a military failure of the "Bay of Pigs" operation to invade Cuba. Cuba did not at this time have the strong relationship with the Soviets that it developed later. Internally, Americans perceived Gagarin's triumph of April 12th 1961 more as an American failure than a Soviet success. As Gagarin's achievements were celebrated around the world, Kennedy contemplated the idea of a face-to-face meeting with the Soviet premier Khrushchev.

When the American ambassador in Moscow in late April 1961 floated the idea of a face-to-face meeting, a declassified top-secret memo[21] reported that Khrushchev had "*shown interest in what he considered to be an American initiative*". Kennedy was already working from a position of weakness; a snub would further undermine his credibility, nationally and internationally. Negotiations over the next few days emphasised the need for secrecy, but discussions over West Germany required Kennedy to consult his partners in France, UK and West Germany. Kennedy had already planned a visit to France and UK in early June. The same memo requested some flexibility in those dates for the London visit as a potential cover if the meeting with Khrushchev did not work out.

Despite the intense secrecy, the 16th May 1961 issue of the *Evening Standard* in London reported that negotiations for a meeting between Khrushchev and Kennedy had been going on for some time. The report did not publish a date, but a two-day summit on 3rd and 4th June in Vienna had, by this time, been

8. A Smile that Changed the World?

agreed. Kennedy would meet with General de Gaulle on the 2nd, then conduct two days of informal talks with Khrushchev in Geneva, and then meet Macmillan in London on the 5th June. These were talks and not negotiations, Kennedy assured his European partners. The main purpose of this face-to-face meeting was to allow each one to get a measure of the other.

The talks would cover Berlin, nuclear testing, disarmament and the potential for USA-USSR scientific collaboration. Suggestions for the scientific collaboration included the sharing of ground based satellite tracking and communication facilities and the co-ordination of major manned space programmes and interplanetary probes.[22] A summary of the two-day talks was provided at a NATO meeting in Paris on the 6th of June.[23] During the talks over approximately eight hours, Khrushchev spoke of Gagarin's spaceflight, saying:

"The principle unknown factor had been the effect of spaceflight on the power of the astronaut to retain the control [of his spacecraft]. Gagarin had to access [the spacecraft] controls through a code, which he would not have been [able to do] had he become irrational, in which case automatic controls would have taken over. In fact, he remained fully rational".

Kennedy spoke strongly of his desire for a nuclear test treaty but against the principle of Troika demanded by Khrushchev. A treaty would require a body that would through a physical inspection enforce the treaty. The Troika principle asserted that the members of the control body should consist of a third representing the West, another third representing the Communist bloc, and the final third represented by the neutral states. Since Khrushchev insisted on unanimity from the controlling body before decisions were implemented, Kennedy concluded this amounted to a veto and could not agree. Khrushchev insisted on it.

In the absence of an agreement, instead of reducing their

8. A Smile that Changed the World?

military presence in Germany, a month later both sides bolstered their forces.[24] On 31st August 1961, Khrushchev announced that the Soviet Union was abandoning its participation in the de facto nuclear testing moratorium that had prevailed since 1958. On 30th October 1961, the Soviets tested the Tsar Bomb, the largest explosion in the history of mankind. A record 140 nuclear tests were conducted in 1962 by the then nuclear powers, UK, USA, France and USSR. That record still stands.

Figure 25 Valentina Tereshkova, Secretary-General U Thant, Yuri Gagarin (16 October 1963 UN) (Courtesy United Nations)

Whilst Gagarin continued charming the people of the world through his world tour of 1961, which some referred to as his "second orbit", tensions over Berlin grew. Potential war was averted as a stalemate emerged when the Soviets built a wall around East Berlin, in August 1961. The Berlin Wall, an icon of the Cold War, remained standing until its spectacular downfall on 9th November 1989.

Peace Envoy

As a Soviet air force Major, Gagarin was more aware than most of the deterioration of the east-west relationship and how close the two super powers edged to an all-out thermonuclear war. He

8. A Smile that Changed the World?

understood from personal experience the violent brutality of war. The German army had arrived at his hometown in 1941 and brought *"an abrupt end to the innocence of [his] childhood"*.[25] He saw things that children should not see and he never forgot them. His brother and sister, taken for two years as slave labourers by the retreating German army, spawned his desire to protect his country through military service. His unique achievement had given him global fame; he toured the world expressing his concerns, hopes and ambitions. As a Major serving in the Soviet air force, he was obliged to follow orders but not as he had done in the past. Having circled the world and constantly feted for it, he seemed to absorb the super human characteristics his audience imagined. In Gagarin's words and demeanour, as a twentieth century Atlas, he assumed the responsibility for world peace. With a large and eager audience, he incessantly and repeatedly extolled his personal and sincere message of peace.

Gagarin raised the topics of peace, collaboration and east west co-operation at every opportunity. At public speeches and press conferences, he repeatedly demonstrated his preoccupation with that theme. In the Academy of Sciences in Moscow, during the press conference in Earl's Court in London, in the AUFW HQ in Manchester and from the back of a lorry in a car park at Trafford Park he preached his singular message on the crucial significance of peace. On the first New Year's Eve in his new life as the world's first spaceman, he sent a new year's greeting to the people of London with his only wish, *"May this be a year of peace on Earth and may the friendship between the British and Soviet peoples develop and grow stronger"*.[26] With an almost tedious persistence, he would repeat the sentiment a few months later.

In his Radio Moscow address (full text in the appendix) in 1962 to the Manchester-based union (AUFW) that had hosted his visit, he referred to the inscription on the medal they had awarded him *"Together moulding a better world"* and went on to

8. A Smile that Changed the World?

say, *"This expresses my own feelings and the feelings of everyone in the Soviet Union. I am sure this is also the wish of the men and women I met in both Manchester and London. Let us work together to solve the many problems facing humanity."*[27]

Gagarin's charisma and gentle charm had won the hearts of the people throughout the UK. His simple smile had beguiled and captivated those that met him. His message of hope and optimism engaged and fascinated the millions who read about him in the papers, heard him on the radio and saw the Richard Dimbleby BBC TV interview during his final evening in the UK. For many British people, Gagarin was the only Russian they had ever known.

The significance of his visit was openly debated in the UK press, an opportunity the Soviet government denied its own people. The Macmillan government was rebuked by many in the letters pages of the newspapers for the initial underwhelming welcome for the Soviet cosmonaut. On the other hand, despite the crowds lining the road to see the cosmonaut drive by in London and Manchester, and the packed halls where he made public appearances, there were also many who cautioned against this excessive celebration of an achievement made by a citizen of a communist country with which Britain could soon be at war.

Gagarin's visit had "*been misunderstood*", wrote the diplomatic correspondent of *The Times* on page 9 of its edition on Monday 19th July, two days after Gagarin left. Reflecting on the tone of the messages sent out by the politicians, the piece makes the point that the warm welcome shown to the cosmonaut should not be considered tantamount to appeasing the Soviets' military crackdown in Berlin. It was nothing more than the expression of admiration of a brave achievement, the correspondent concluded.

The Russian communist paper *Pravda* put its own spin on Gagarin's UK visit by claiming that the visit had "*become a*

8. A Smile that Changed the World?

political issue for the whole of the country".[28] It claimed that the British people were "*fed-up*" of the Cold War. This was at least in part true. Some letters in the British newspapers warned against the dangers of being seduced by Soviet propaganda. In response others wrote back emphasising the importance of communication between the peoples, especially at times when the word "war" had entered the politicians' vocabulary:

"*It is precisely because we are aware of the hideous dangers confronting us that some of us take the opportunity of "getting through" to the Russian people. We are aware that the communist propaganda distorts our every word and action and the massive veto on communication stands between us. Nevertheless, we refuse to believe that if they knew the truth, the Russian people would be any more thirsty for our blood than we are for theirs. We must lie or die together. What, then, can we do but grasp the all too rare occasions for personal contact?*"[29]

The reaction of the West German people (who never experienced first-hand his charm) seemed less divided, as conveyed in a confidential memo from Bonn to the Foreign Office on 15th July. It expressed the alarm and dismay of ordinary Germans over the reception given by the "*reserved and politically mature British people to a Bolshevik on a propaganda mission*".[30]

Fifty years after meeting Gagarin, Stanley Nelson, who then was working as an illustrator in the drawing office at Metrovicks, remembered that Gagarin was the only Russian he had ever seen. Although he potentially was a representative of a country with which war could break out at any time, Gagarin's smiling face had brought a smile to the thousands present at the Manchester based Metrovicks in Trafford Park. Nelson recalled that Gagarin's "Russianness" was irrelevant. "*The man on the street doesn't mind, he doesn't get involved and accepts people for who they are.*"[31]

8. A Smile that Changed the World?

Figure 26 Yuri Gagarin and Albert Knight, Moscow, April 1963
(Courtesy Cath Todd)

Two years after Gagarin's visit to Manchester, Albert Knight, an 86-year-old from Rochdale in Lancashire, embarked on a peace mission of his own and met Gagarin in Moscow. Knight left Rochdale on 22nd April 1963 for Tilbury Docks. From there he and other eleven travelled to Moscow by steamships and overnight sleepers. The group, sponsored by the Communist Party of Great Britain (CPGB), toured the USSR during a six week visit, experienced the May Day parade in Red Square and had tea with Yuri Gagarin. Knight had been one of the earliest members of the CPGB. If Gagarin acquired his moral values from the trauma of the Second World War, Knight acquired his on the battlefields of the first.

On his return to Rochdale, Knight insisted that *"Russia wants to live at peace with the world. They are like the British people in their outlook. They don't want war"*.[32] Knight and the other eleven members of the CPGB probably did not need to visit

8. A Smile that Changed the World?

Moscow to hold that view, but he was a member of a tiny minority in the UK who had first-hand experience of the Soviet Union from the inside.

Figure 27 Painting by Walter Kershaw to commemorate the 50th anniversary of Gagarin's visit to Manchester. (Courtesy Walter Kershaw)

The narrow perceptions of socialism, the Soviet Union and its people held by the British people were largely products of ignorance brought about by the closed nature of the Soviet society. The lack of a free press and oppressive nature of the Soviet secret services prevented the Soviet people from understanding their own society as well as those outside. In a launch-pad explosion in October 1960, over one hundred people died. That incident remained a state secret for almost three decades. If Gagarin had been killed during launch, that event too would have been swept under the carpet of state secrets[33] and the world would have remained oblivious to the name of Yuri Gagarin.

8. A Smile that Changed the World?

Manchester marked the fortieth anniversary of Gagarin's flight with an exhibition at the Museum of Science and Industry (MOSI) and at Jodrell Bank.[34] Perhaps the existence of what became to be popularly known as "the space race" had itself averted war. Present at the exhibition at the MOSI, Don Lind, Space Shuttle astronaut, said in 2001, *"The space race was really a substitute for war. We made great technological advances without shooting anyone"*.[35]

On 12th April 2011, events throughout the UK and the world celebrated the fiftieth anniversary of this unique achievement and the man who risked his life to accomplish it.36 The Soviet Union marked the first anniversary on 12th April 1962 with a national holiday and that tradition continues today. The legacy of Gagarin's memories lingers, not only in the recollections of those who saw and met him, although not many of them remain, but also in the intangible hopes and aspirations of the "better world" inscribed in the medal he was presented with by the working people of Manchester.

What was the impact of Gagarin's five days in Britain? Was his quest for peace, friendship and greater international collaboration successful? A quantitative answer is impossible, but qualitatively the testimony of those that met him and the debate in the immediate aftermath of his visit demonstrates it was. For many he did change their perception of the Soviet Union and through that inspired a more hopeful future. At a critical time of fragile international relationships, Gagarin's presence in Britain provided a rare but critical opportunity to peer inside the blue eyes of a handsome and unpretentious farm boy, a representative of the until then unseen enemy.

Events combine in complex and unexpected ways to deliver unexpected outcomes. Bernard Lovell became an astronomer motivated to understand transient radar echoes that should have not been there.[37] Buzz Aldrin failed twice in his application for a Rhodes Scholarship but succeeded in one of humanity's greatest

8. A Smile that Changed the World?

adventures.[38]

Gagarin's persistence over the five days had cultivated desire for peace in the British people that filtered through them to their political leaders. At a time when politicians were contemplating decisions on war and peace, Gagarin's presence brought unexpected optimism and hope. Gagarin merged his exceptional accomplishment with his ordinariness and communicated with the masses instinctively through his smile. His unparalleled achievement and the popularity that followed gave his message a reach in distance and depth that the world needed during the coldest days of the Cold War.

Appendix

Glossary

AEEU	Amalgamated Engineering and Electrical Union. The Union where the AUFW was amalgamated in 1968
AEI	Associated Electrical Industries
Astronaut	A person who travels in space (US). Cosmonaut in Russia
Astronautics	The science of space flight
AUFW	Amalgamated Union of Foundry Workers
BIS	British Interplanetary Society
BSFS	British Soviet Friendship Society. Sometimes referred to as the Anglo Soviet Friendship Society
BSFS	British Soviet Friendship Society
CPGB	Communist Party of Great Britain
CPUSA	Communist Party of the USA
Ejection seat	Pilot seat fitted with explosive charge for emergency exit from a space or aircraft
Escape velocity	A velocity required to overcome the pull of gravity. The velocity necessary to escape from the Earth is 11.2 km/second
FIMS	Friendly Iron Moulders' Society
FSIF	Friendly Society of Iron Founders
GB-USSR	Great Britain – Union of Soviet Socialist Republics A HMG sponsored organisation. Today called the GB-Russia Society
Heat shield	A device which protects people or equipment from heat, such as a shield in front of a re-entry capsule
HMG	Her Majesty's Government
IAF	International Aeronautical Federations
ICBM	Intercontinental Ballistic Missile. The R7 is credited with being the first tested successfully in 1957

IGY	International Geophysical Year (1957-58)
Izvestia	A high circulation daily newspaper in Russia. Up until 1991, it was the official Soviet Government communication. The word means "delivered message"
Kerosene	A petroleum based rocket fuel
LAC	Lancashire Aero Club
LEO	Low Earth Orbit (typically between 90 and 200 kilometres)
Metrovicks	Metropolitan Vickers Electrical Company
NASA	National Aeronautics and Space Administration
Pravda	High circulation newspaper. Up until 1991 it was official communication of the Communist Party in the USSR. A tabloid with the same name, independent of government control, exists today.
RICS	Royal Institution of Chartered Surveyors. One of the several organisations associated with extending a visit to Gagarin
Satellite	Natural or artificial, in orbit around a planet, a Moon or planet
SCRSS	Society for Cultural Relations and Soviet Studies. Today known as "The Society for Co-operation in Russian and Soviet Studies" and frequently abbreviated to SCR.
Solid propellant	A rocket propellant in solid form. Once ignited it cannot be stopped
Solid rocket booster	A rocket, usually powered by solid propellants, to assist a spacecraft to reach orbit
Space	The point beyond the Earth's atmosphere. International standards have established that space begins at an altitude of 100 km
Sputnik	The world's first satellite launched on 4th October 1957 by the USSR. The word means "companion" or "satellite".
TASS	A long established, state controlled, Soviet news agency. Had a high profile during the Soviet era
Trajectory	The path taken by a projectile, rocket or satellite
Voskhod	The spacecraft that followed Vostok. The word Voskhod means "Accent" or "Dawn".
Vostok	Name of Gagarin's spaceship. The word means "East" in Russian
Weightlessness	A prolonged state of free fall where weight is absent

Gagarin's Radio Message to British Foundry Workers

The text of a message broadcast by Yuri Gagarin in April 1962, the first anniversary of his spaceflight on Radio Moscow. This transcript is from the Journal of the Amalgamated Union of Foundry Workers 1962.

Dear Brothers! I am very glad of this opportunity to speak to you on a day so memorable to me. I remember my arrival in Manchester on a rainy morning in July and how warmly I was welcomed by the leaders of your Union, by the Lord Mayor, and by the people of Manchester.

This was followed by an event which moved me deeply: British foundry workers, men in what was once my own trade, accepted me into their militant Union. I take this opportunity to convey my thanks for this honour once again to Fred Hollingsworth, the president of the Amalgamated Union of Foundry Workers, David Lambert the general secretary and all the members of the Union.

I also thank the Union leadership for their warm telegram of greetings on this anniversary. It's a long time since I was a moulder but I haven't forgotten the trade. I'm proud to belong to the working class, which brought me up and sent me on the road to outer space.

I remember the huge yard of the Metropolitan Vickers works where I was able to speak to the workers. And although the sky was overcast and there was even a drizzle, I was warmed by the friendly smiles of the thousands of people in working togs who filled the yard.

And the firm handshakes of my fellow workers in the moulding shop were dearer to me than many awards. At the meeting on that occasion, I said that I was boundlessly glad of the opportunity to shake thousands of working hands, the hands that create everything beautiful on Earth.

I shake your hands again, dear brothers, and wish you success in your life and work! Inscribed on the gold medal which you gave me in Manchester are the words: "Together we shall mould a better world!"

This expresses my own feelings and the feelings of everyone in the Soviet Union. I am sure this is also the wish of the men and women I met in both Manchester and London.

Let us work together to solve the many problems facing humanity.

I know that there is a traditional bond of friendship between your Union and the Soviet engineering union to which I, too, belonged when I worked at a factory. I am very glad that this bond exists, and I am confident that this friendship will grow from year to year. As for my personal plans, I am working hard and studying. Sometimes I also have to act as teacher, passing on my space experience to our future spacemen.

Hearty greetings to your families from my wife, Valentina, and from my daughters, Galya and Lena! And all the best from me!

Gagarin in Britain: Timeline

Tuesday 11th July
10:30 Arrival at Heathrow [*Guardian* 11/7/1961]
11:45 Soviet Embassy [*Daily Worker* 10/7/1961]
13:00 Earl's Court [*Guardian* 12/7/1961]
15:00 Press conference in Fashion Hall Earl's Court [*Guardian* 12/7/1961]
16:15 BIS medal award at the end of the press conference [*Flight* 20/7/1961]
16:30 Leave Earl's Court [*Guardian* 12/7/1961]
17:30 Evening reception at Soviet Embassy [*Daily Worker* 10/7/1961]

Wednesday 12th July
10:00 Arrival at airport [*Manchester Evening News* 11/07/1961]
10:45 AUFW Medal Ceremony [*Guardian* 12/7/1961]
11:35 Metropolitan-Vickers at Trafford Park [*Manchester Evening News* 11/7/1961]
12:45 Manchester Town Hall [*Manchester Evening News* 11/7/1961]
16:30 Manchester Airport

Thursday 13th July
11:00 Mansion House - Lord Mayor of London [*Daily Worker* 13/7/1961]
11:45 Tower of London - Gv. Sir Thomas Butler [*Daily Worker* 13/7/1961]
13:20 Burlington House - Royal Society
15:00 Return to USSR Embassy
15:45 Meet PM at Admiralty House [Prem 11-3543 12/07/1961 National Records Archive]
16:15 Lays wreath at the cenotaph
16:30 Air Ministry in Whitehall - Secretary of State for Air

[*Daily Worker* 13/7/1961]
18:00 Hyde Park Hotel GB USSR Association [*Daily Worker* 13/7/1961]
19:30 Muscovites-Association cancelled. Sightseeing tour instead [*Guardian* 14/7/1961]
22:15 Back at USSR Embassy [*Daily Worker* 13/7/1961]

Friday 14th July
12:50 Buckingham Palace [FO 371-159606 12/07/1961 National Records Archive]
14:45 Soviet Embassy
15:30 Earl's Court [*Daily Worker* 14/7/1961]
16:00 Highgate Cemetery [Time is uncertain; *The Times* 15/7/1961 says "Evening"]
16:40 Soviet Embassy – British Soviet Friendship Society [*Daily Worker* 14/7/1961]
21:00 Earl's Court Live BBC TV interview (at 21:30) from Earl's Court with Richard Dimbleby, Tom Margerison, science editor of the *Sunday Times*, and Yuri Fokin of the Soviet Television Service [*Daily Worker* 14/7/1961]
22:15 Soviet Embassy

Saturday 15th July
11:00 Leave Soviet Embassy for Airport [*Daily Worker* 15/07/61]
11:45 Press conference at Airport
12:35 Depart for Soviet Union [*Daily Worker* 15/07/61]
18:05 Arrive in Moscow [FO 371-159606 12/07/1961 National Records Archive]

Bibliography: Books Cited

AUFW pamphlet: "Welcome Brother Gagarin: A Pictorial Record of a Great Day". Manchester 1961.

Bizony, P. & J. Doran: *Starman: The Truth Behind the Legend of Yuri Gagarin.* New York 1998.

Brown, Captain E.M.: *Wings on My Sleeve.* Shrewsbury 1961 & 1978.

Burchett, W. & A. Purdy: *Cosmonaut Yuri Gagarin: First Man in Space.* Norwich 1961.

Burgess, C. & F. French: *Into that Silent Sea: Trailblazers of the Space Era, 1961-1965.* Lincoln 2007.

Burgess, C. & R. Hall: *The First Soviet Cosmonaut Team: Their Lives, Legacy and Historical Impact.* Berlin 2009.

Chertok, B.E.: *Rockets and People*, Volume 1, NASA SP-2005-4110. Washington DC 2005.

Engels, F.: *Conditions of the Working Class in England: With a Preface Written in 1892.* Transl. F.K. Wischnewetzky. London 1892.

Evans, B: *Escaping the Bonds of Earth: The Fifties and the Sixties.* Berlin 2009.

Fryth, H.J. & M. Collins: *The Foundry Workers: A Trade Union History.* Manchester 1959.

Gagarin, Y. et al.: *Soviet Man in Space.* Honolulu 1961.

Gagarin, Y.: *Road to the Stars: Notes by Soviet Cosmonaut No. 1.* Told to N. Denisov and S. Borzenko. Ed. N. Kamanin. Moscow ca. 1961.

Golovanov, IA.K.: *Sergei Korolev: The Apprenticeship of a Space Pioneer.* Moscow 1975.

Gray, T.: *Manchester Ship Canal. Stroud* 1997.

Harford, J.: *Korolev: How One Man Masterminded the Soviet Drive to Beat America to the Moon.* New York 1997.

Jenks, A.L.: *The Cosmonaut Who Couldn't Stop Smiling: The Life and Legend of Yuri Gagarin.* DeKalb IL. Forthcoming July 2011.

Lovell, B: *Astronomer by Chance.* Basingstoke 1991.

Moore, P: *Eighty Not Out: The Autobiography.* London 2003.

Oberg, J.E.: *Red Star in Orbit: The Inside Story of Soviet Failures and Triumphs in Space.* New York 1981.

Ollerenshaw, K.: *First Citizen.* London 1977.

Rynin, N.A.: *Interplanetary Flight and Communication.* NASA Technical Translation F-642 to 648. Washington DC 1971.

Scott, D. & A. Leonov: *Two Sides of the Moon: Our Story of the Cold War Space Race.* New York 2004.

Siddiqi, A.A.: *Challenge to Apollo: The Soviet Union and the Space Race: 1945 – 1974.* Washington DC 2000.

Spiers, M.: *Victoria Park Manchester.* Manchester 1976.

Tsiolkovsky, K.E.: *Works on Rocket Technology.* Moscow, 1947. NASA Technical Translation F-243. Washington DC 1965.

Tsymbal, N.A.: *First Man in Space: The Life and Achievement of Yuri Gagarin: A Collection.* Moscow 1984.

Turnill, R.: *The Moon Landings: An Eye Witness Account.* Cambridge 2003.

Ward, B.: *From Nazis to Nasa: The Life of Wernher von Braun.* Stroud 2006.

Index

Admiralty House 92, 99, 100, 104, 105
Air Ministry 101
Alcock, John 51, 52, 53, 54
Aldrin, Buzz 19, 41, 55, 146
Amalgamated Union of Foundry Workers. 21, 44, 45, 51, 56, 62, 63, 64, 66, 67, 69, 75, 77, 80, 83, 90, 91, 99, 126, 141, 150
  cine film 66, 69, 87
  gold medal 64, 126, 151
Amery, Julian 34, 108
Anglo-Soviet Friendship Society 31
Anniversary celebrations 146
Apollo 128
Apollo 11 .... 12, 41
Apollo 16 96
Apollo 8 .... 18, 85
Apollo-Soyuz Test Project 16
Armstrong, Neil ... 19, 41, 55, 127, 128
Avro 504K 1, 54
Avro Transport Company 54
Belitzky, Boris.. 27, 28, 29, 33, 34, 38, 45, 66, 78, 94, 97, 100, 104, 105, 116, 117, 120, 122, 125, 131
Biggs, Lionel 47
Blagnarovov, Anatoli 132
Borisov, Boris 20
Braun, Wernher von . 8, 13, 14, 15, 16, 17, 18, 41, 104, 128, 136
Brewster, James 34, 42
Briggs, Lionel 44
British Interplanetary Society .. 23, 39, 40, 41, 42, 99
British Soviet Friendship Society 23, 121
British Trade and Industrial Exhibition 25
British Trade Fair 26
Brown, Arthur Whitten 53
Brown, Eric 102, 103, 104, 105, 106, 107, 108
Buckingham Palace 21, 28, 29, 91, 92, 111, 116, 117
Buckler, Philip 114, 115, 116
Chertok, Boris 128
Clarke, Arthur C 13, 30, 40
Cold War 22, 56, 85, 87, 136, 137, 138, 139, 140, 147
Communist Party of Great Britain 20, 31, 43, 65, 144
Cutlery fail 119
Denisov, Nicolay 27, 100, 116
Dimbleby, Richard . 122, 123, 129, 142
Discovery Class twin-engine turbo-prop BEA Viscount 800 45
Dog 36
Dornberger, Walter 14
Endert, Hans 17
Engels, Friedrich 74, 111
Fokin, Yuri 122
Foundry work 58, 59, 60
French, Francis 87
Gagarin, Yuri Alekseyevich ... 129, 140, 142, 143, 144, 150
  arrival in UK 19
  awards and presents 28, 32, 39, 40, 63, 66, 67, 98, 100, 108, 121
  death 54, 62, 134, 135, 136
  early life 37, 57, 58, 73, 85, 104, 118, 120, 140
  education 1
  ejected from spacecraft ... 7, 35, 36, 101, 107, 132

family 27, 31, 38, 57, 119, 134, 141, 151
first space flight... 3, 4, 5, 6, 19, 35, 37, 83, 124, 128, 129, 130, 133, 138, 139, 145
foundryman... 2, 44, 45, 57, 58, 59, 67, 73, 76, 78, 150
personality... 22, 30, 37, 38, 39, 44, 46, 48, 74, 78, 79, 87, 89, 97, 100, 101, 105, 114, 119, 122, 131, 134, 136, 141, 142, 144
pilot 45, 46, 102, 134, 135
selection 2, 28
significance 19, 30, 56, 57, 130, 133, 138, 143, 145, 146
world tour............. 19, 140, 141
GB-USSR Association 23, 24, 108, 109, 110
Goddard, Robert.............. 7, 15, 16
Hero of the Soviet Union... 28, 38, 66
Highgate Cemetery 111, 113
Hochhouser, Victor & Lillian... 24, 25
Hollingsworth, Fred 45, 47, 64, 65, 66, 67, 90, 91, 150
Ilyushin-18 129
International Aeronautical Federation 35, 107, 108
International Geophysical Year 127
Invitation to UK 20, 21, 22, 23, 24, 25, 26, 27, 45, 75
Jodrell Bank 21, 50, 84, 85, 86, 87, 91, 146
Jolly, Gilbert.............................. 26
Kamanin, Nikolai... 27, 28, 33, 45, 97, 100, 101, 107, 114, 116, 125
Kennedy, John F... 22, 33, 41, 137, 138, 139
Kershaw, Walter 145
Khrushchev, Nikita Sergeyevich 2, 25, 26, 33, 37, 57, 100, 127, 129, 130, 138, 139, 140
Kibalchich, Nikolai..... 8, 9, 10, 14

Knight, Albert 144
Komarov, Vladimir 134
Korolev, Sergei Pavlovich.. 1, 7, 8, 12, 15, 18, 119, 127, 128, 129, 130
Kuznetsov, Igor 136
Lambert, David 65, 150
Leonov, Alexi..... 16, 39, 134, 135, 136
Lind, Don 146
Lloyd, Alf.......... 63, 64, 69, 75, 88
London 30
cenotaph 101
departure 125
eyewitnesses....... 33, 34, 95, 96
itinerary 93
sightseeing.......... 62, 92, 95, 96
Lovell, Bernard 12, 50, 83, 84, 85, 86, 87, 88, 97, 128, 146
Luna 3 .... 85, 86, 87
MacLean, Fitzroy.............. 23, 108
Macmillan, Harold . 21, 24, 28, 32, 44, 91, 92, 93, 99, 100, 101, 102, 108, 109, 110, 142
Manchester............................... 44
arrival 44, 46
cemetery.............................. 51
cenotaph 80
eyewitnesses 48, 49, 54, 55, 67, 68, 75, 76, 78, 79, 80, 81, 88, 89, 90, 143
industry 57, 59, 70, 71, 74
itinerary . 45, 48, 50, 51, 54, 55, 56, 65, 75, 77, 81, 88, 89
significance for airplane industry 1, 52, 53, 54
Manchester Airport 44, 46, 47, 51, 52, 90
Manchester Town Hall . 49, 51, 80, 81, 82, 88
reception............. 83, 86, 87, 97
Mansion House 92, 93, 94
Margerison, Tom............. 122, 123
Marx, Karl........... 72, 74, 111, 112
Mason, R.H. 23, 100
Maudling, Reginald............. 26, 43

Metrovicks .26, 69, 72, 73, 75, 76, 143, 149, 150
MiG 129
MiG-15 .... 134, 135, 136
Moore, Patrick 40, 87, 101
Muscovites Association 110
NASA 15, 37, 42
Nelson, Stanley 75, 78, 143
Nikolayev, Andrian 134
Oberth, Hermann. 7, 13, 14, 15, 16
Ollerenshaw, Kathleen 50, 51
Peenemünde 16, 17
Polycarp R-5 biplane 28
Poyekhali 3
Press conference
 London 35, 37, 38, 42, 141
 Moscow 35, 130, 131, 132, 133, 141
Queen Elizabeth II 44, 93, 94, 117, 118, 119, 121
R7 3, 4, 127, 128
Radio Moscow address to AUFW
 .... 141, 150
Rochdale .... viii, 82, 144
Roe, Alliot Verdon 1, 52
Royal Society 92, 97, 98, 99
 reception 97, 98
Saratov 7, 39, 58
Sausages 119
Seregin, Vladimir 134, 135
Shepard, Alan 23, 37, 107
Society for Cultural Relations ... 31
Soldatov, Alexander 27, 33, 44, 45, 47, 97, 100, 116, 125
Soviet Embassy 22, 23, 28, 30, 32, 34, 42, 92, 99, 109, 124, 126
 reception 43
Soviet space programme .. 1, 2, 12, 21, 86
Soviet Trade Fair 21, 22, 26, 28, 29, 30, 34, 42, 61, 77, 122
Space cooperation USA and USSR 16
Space race 18, 36, 128, 146
Sputnik 12, 42, 83, 84, 85, 115, 127, 128, 129, 130
Stafford, Tom 16
SU-15 .... 135, 136
Tereshkova, Valentina 37, 41, 134, 140
Titov, Gherman 19, 35, 36
Tower of London 92, 94, 95, 96
Trade unionism 60, 61, 62
 foundrymen's unions 61, 62
Trafford Park 72, 77, 80
Tsiolkovsky, Konstantin ... 7, 8, 10, 11, 12, 13, 15, 16, 127, 128
Tupolev TU104B 32
Turnbull, Francis 32, 45, 125
Turnill, Reg .. 17, 18, 35, 101, 130, 131, 132
Utochkin, Sergei 1
V2 16, 17, 41, 136
Vimy IV twin-engine biplane 53
Voskhod 36
Vostok 3, 5, 6, 11, 36, 37, 83, 129, 130, 133
Waley-Cohen, Bernhard Nathaniel
 .... 92, 93, 94
Zarnecki, John 115, 116

Notes and References

1. Space Age

1 Golovanov, IA.K.: Sergei Korolev: The Apprenticeship of a Space Pioneer, 1975 p30.
2 Ibid, p188.
3 Ibid, p166.
4 Jenks, A.: The Cosmonaut Who Couldn't Stop Smiling: The Life and Legend of Yuri Gagarin, forthcoming July 2011.
5 Burgess, C. & R. Hall: The First Soviet Cosmonaut Team, 2009, p45.
6 Ibid, p20.
7 Extracts taken from the Audio files provided by the Russian State Archive for the film "First Orbit" in April 2011 and translated to English by Chris Riley, Iya Whiteley and Stephen Slater.
8 Ibid.
9 Ibid.
10 Ibid.
11 Bizony, P. & J. Doran: Starman: The Truth behind the Legend of Yuri Gagarin, 1998, p103.
12 Scott, D. & A. Leonov: Two Sides of the Moon, 2004, p297.
13 Extracts taken from the Audio files provided by the Russian State Archive for the film "First Orbit" in April 2011 and translated to English by Chris Riley, Iya Whiteley and Stephen Slater.
14 Harford, J.: Korolev, 1997, p9.
15 Rynin, N.A.: Interplanetary Flight and Communication, Volume 2 number 4, NASA Technical Translation F-646, 1971, p36.
16 Ibid.
17 Ibid.
18 Tsiolkovsky's first work on rockets, "Rockets in Cosmic Space," was published in 1903 in the journal Number 5 Nauchnoye obozreniy" with the different title "Investigation of Universal Space by Means of Reactive Devices".
19 Tsiolkovsky, K.E.: Works on Rocket Technology. Moscow, 1947, NASA Technical Translation F-243, 1965.
20 Ibid.
21 Ibid.
22 Ibid.
23 Ward, B.: From Nazis to Nasa, 2006, p227.
24 Ibid, p17.
25 Tsiolkovsky, K.E.: Works on Rocket Technology. Moscow, 1947, NASA Technical Translation F-243, 1965, p7.

26 Ibid.
27 Chertok, B.E.: Rockets and People, Volume 1, NASA SP-2005-4110, p243.
28 Von Braun's 161 page PhD thesis, entitled "Construction, Theoretical, and Experimental Solution to the Problem of the Liquid Propellant Rocket" was instantly classified as top-secret and not published until 1966. The original copy was sold for £33,000 on 5/12/2007.
29 Harford, J.: Korolev, 1997, p42.
30 http://www.astronautix.com/astros/oberth.htm [01/06/2011].
31 http://www.airspacemag.com/space-exploration/russian_space_dream.html?c=y&page=1 [01/03/2011].
32 Ibid.
33 http://history.nasa.gov/SP-4209/ch11-3.htm [03/03/2011].
34 Chertok, B.E.: Rockets and People, Volume 1, NASA SP-2005-4110, p250.
35 Turnill, R.: The Moon Landings: An Eye Witness Account, 2003, p2.
36 Chertok, B.E.: Rockets and People, Volume 1, NASA SP-2005-4110, p306.
37 Siddiqi, A.A.: Challenge to Apollo: The Soviet Union and the Space Race, 2000, p687.

2. An Uneasy Invitation

1 Time Magazine 18th August 1961.
2 Tsymbal, N.A.: First Man in Space, 1984, p89.
3 Guardian 30th June 1961.
4 Guardian 8th July 1961.
5 Mason, R.H.: Foreign Office Papers 371-159606, 7th July 1961, National Records Archive.
6 Guardian 11th July 1961.
7 Daily Telegraph 13th July 1961.
8 Mason, R.H.: Foreign Office Papers 371-159606, 16th June 1961, National Records Archive.
9 Mason, R.H.: Foreign Office Papers 371-159606, 11th July 1961, National Records Archive.
10 James, C.: Foreign Office Papers 371-159606, 21st June 1961, National Records Archive.
11 Mason, R.H.: Foreign Office Papers 371-159605, 1st May 1961, National Records Archive.
12 Guardian 24th April 1961.
13 Thomas, D.: Foreign Office Papers 371-159606, 5th May 1961, National Records Archive.
14 Millar, F.: Foreign Office Papers 371-159605, 21st April 1961, National Records Archive.
15 Hansard 27th April 1961.
16 Ibid.

17 Oberg, J.E.: Red Star in Orbit, 1981, p. 97.
18 Evans, B.: Escaping the Bonds of Earth, 2009, p6.
19 http://english.ruvr.ru/radio_broadcast/2248140/ [30/01/2011]
20 http://www.foia.cia.gov/search.asp?pageNumber=1&freqReqRecord=GaryPowers.txt [01/02/2011].

3. Like Some New Columbus

1 Clarke, A.C.: "Profiles of the Future", 1961.
2 Elena Gagarina interviewed by Andrea Rose. 11th April 2011.
3 Daily Worker 12th July 1961.
4 Guardian 12th July 1961.
5 Wilson, D.: Foreign Office Papers 371-159607, 21st July 1961, National Records Archive.
6 Guardian 20th May 1961.
7 Email correspondence with the author 14th March 2011.
8 Daily Worker 8th July 1961.
9 Guardian 12th July 1961.
10 Flight Magazine 20th July 1961
11 Interview with the author 19th January 2011.
12 Siddiqi, A.A.: Challenge to Apollo: The Soviet Union and the Space Race, 2000, p283.
13 Time Magazine 18th August 1961.
14 Siddiqi, A.A.: Challenge to Apollo: The Soviet Union and the Space Race, 2000, p267.
15 Flight 20th July 1961.
16 The Times 12th July 1961.
17 Ibid.
18 Gagarin, Y.: Road to the Stars, Told to Nikolay Denisov and Sergei Borzenko, ed. N. Kamanin, ca. 1961, p30.
19 Scott, D. & A. Leonov: Two Sides of the Moon, 2004, p39. However, this cannot be accurate since Hemmingway died in 1961.
20 Flight 20th July 1961.
21 Moore, P: Eighty Not Out, 2003, p21.
22 Journal of the British Interplanetary Society 9, 1950.
23 Burchett, W. & A. Purdy: Cosmonaut Yuri Gagarin: First Man in Space, 1961, p31.
24 President Kennedy's speech to Congress 25th May 1961.
25 Daily Worker 11th July 1961.
26 Spaceflight (BIS Journal), September 1970.
27 The Times 12th July 1961.
28 Daily Worker 12th July 1961.

4. Together Moulding a Better World

1 Guardian 12th July 1961.
2 The Foundry Workers Journal, Monthly Report of the Amalgamated Union of Foundry Workers, May 1961.
3 Ibid.
4 It was broken up in February 1997, and since 5th December 2006 the remaining forward fuselage has been on show at Bournemouth Aviation Museum. http://vickersviscount.net/Index/VickersViscount263History.aspx [21/01/2011].
5 Evening Chronicle 12th July 1961.
6 Ibid.
7 Manchester Evening News 12th July 1961.
8 Manchester Evening News 11th July 1961.
9 Email correspondence with the author 17th November 2010.
10 Email correspondence with the author 6th November 2010.
11 Email correspondence with the author 7th November 2010.
12 Ollerenshaw, K.: First Citizen, 1977, p135.
13 Interview with the author 6th October 2010.
14 Ibid.
15 Email correspondence with the author 30th September 2010.
16 Email correspondence with the author 7th November 2010.
17 Gagarin, Y. et al., Soviet Man in Space, 1961, p20.
18 During the 1930s all privately owned farms had been forced by Stalin to become collective farms.
19 Gagarin, Y.: Road to the Stars, Told to Nikolay Denisov and Sergei Borzenko, ed. N. Kamanin, ca. 1961, p21.
20 Ibid,, p38.
21 Bizony, P. & J. Doran: Starman: The Truth behind the Legend of Yuri Gagarin, 1998, p23.
22 Fryth, H.J. & M. Collins: The Foundry Workers, 1959, p9.
23 Ibid., p16.
24 This act made it illegal for employees to seek better conditions of employment. Masters used the act to imprison employees agitating for improvements in their conditions of employment.
25 Fryth, J. & M. Collins: The Foundry Workers, 1959, p41.
26 Spiers, M.: Victoria Park Manchester, 1976.
27 Guardian 12th July 1961.
28 Journal of the AUFW 1961.
29 Evening Chronicle 12th July 1961.
30 AUFW film from 12th July 1961. North West Film Archive 1961.
31 Guardian 13th July 1961.

32 Manchester Evening News 12th July 1961.
33 An edited version available here [http://astrotalkuk.org/2011/04/12/episode-42-april-12th-2011-rare-video-of-yuri-gagarin-in-manchester/]
34 AUFW film from 12th July 1961. North West Film Archive 1961.
35 Manchester Evening News 12th July 1961.
36 The secret high altitude spy plane was shot down on 1st May 1960 and its pilot taken prisoner. Khrushchev demanded an apology at the Four Powers Paris summit on 16th May 1960. He never got one and stormed out, which led to deep mistrust particularly between the USA and USSR, which probably stoked the fire of the space race.
37 Manchester Evening News 12th July 1961.
38 Daily Telegraph 13th July 1961.
39 Daily Worker 13th July 1961.
40 Email correspondence with the author 1st December 2010.
41 Interview with the author 12th November 2010.
42 Ibid.
43 The AUFW merged in 1968 to become AEEU. Another merger 2001 formed Amicus. The most recent merger in 2007 created Unite where the spirit of AUFW currently resides.
44 Conversation with the author 5th November 2010.
45 During a special viewing of this cine film for Francis French at the union office in 1987 a section broke off. The section was given to Francis French who assisted in digitally reuniting it with the original in 2011. See http://blogs.sandiegoairandspace.org/library/?p=799

5. Working Class Cosmonaut

1 Gray, T.: Manchester Ship Canal, 1997.
2 http://www.modernhistory.co.uk/site/media/facts-and-figures [10/04/2011].
3 Gagarin, Y.: Road to the Stars, Told to Nikolay Denisov and Sergei Borzenko, ed. N. Kamanin, ca. 1961, p22.
4 Ibid.
5 http://www.spinningtheweb.org.uk/ [01/02/2011].
6 Engels, F.: Conditions of the Working Class in England, 1844, p4.
7 Conversation with the author 27th March 2011.
8 Ibid.
9 See French, F. in an article about Gagarin's visit to Manchester published in British Interplanetary Society's magazine Spaceflight in July 1987. Also available here: http://yurigagarin50.org/history/gagarin-in-britain/gagarin-in-manchester.
10 AUFW pamphlet: "Welcome Brother Gagarin: A Pictorial Record of a Great Day", 1961.
11 Journal of the Amalgamated Union of Foundry Workers 1961.

12 Full transcript of his radio message from April 12th 1962 in appendix.
13 Email correspondence with the author 22nd November 2011.
14 Conversation with the author 18th December 2010.
15 Manchester City Council adopted the symbol of the hard working bee for the city to reflect the industriousness of its people.
16 Evening Chronicle 12th July 1961.
17 The Soviets concealed their plans of further Sputnik launches by not giving the first Sputnik a suffix. It was known originally as Sputnik and only later as Sputnik 1. This was repeated with Vostok.
18 Lovell, B: Astronomer by Chance, 1991, p259.
19 Ibid.
20 Ibid. p262.
21 Guardian 28th December 2008.
22 Email correspondence with the author 28th March 2011.
23 Conversation with the author, 26th January 2011.
24 Trafford Today: The Newspaper of Trafford Council, January 2009.
25 Journal of the Amalgamated Union of Foundry Workers, 1961.

6. A Delightful Fellow

26 Daily Worker 14th July 1961.
27 Email correspondence with the author 26th October 2010.
28 Telephone interview with the author 11th January 2010.
29 Ibid.
30 Guardian 14th July 1961.
31 Notes and Records of the Royal Society 1st November 1961.
32 First: Planets have elliptical not circular orbits. Second: they move the fastest when nearest the Sun. Third: time taken for a complete orbit is proportional to the distance from the Sun.
33 The source http://royalsociety.org/about-us/fellowship/honorary-fellows/ states seven but lists six (Melvyn Bragg, David Sainsbury, Nora O'Neill, Sir Ralph Kohn, Leonard Wolfson, Sir Patrick Moore) [01/06/2011].
34 Evans, S.: Prem 11-3543, 11th July 1961, National Records Archive.
35 Evans, S.: Prem 11-3543, 13th July 1961, National Records Archive.
36 The Times 14th July 1961.
37 Moore, P: Eighty Not Out, 2003, p48.
38 Interview with the author 19th January 2011.
39 Daily Worker 14th July 1961.
40 See http://www.flightglobal.com/airspace/forums/captain-eric-winkle-brown-greatest-7922.aspx [01/02/2011].
41 Brown, E.: Wings on My Sleeve, 1961 & 1978, p71.
42 Interview with the author 19th January 2011.
43 Ibid.

44 Ibid.
45 Telephone conversation with the author 14th March 2011.
46 Bizony, P. & J. Doran: Starman: The Truth behind the Legend of Yuri Gagarin, 1998, p 137.
47 Kamanin Diaries, http://www.astronautix.com/articles/kamdiaries.htm [10/03/2011].
48 Guardian 14th July 1961.
49 49 Mason, R.H.: Foreign Office Papers 371-159606, 11th July1961 National Records Archive.
50 Bligh, T.: Prem 11-3543, 11th July1961, National Records Archive.
51 Guardian 14th July 1961.

7. Communism and Royalty

1 Gagarin, Y.: Road to the Stars, Told to Nikolay Denisov and Sergei Borzenko, ed. N. Kamanin, ca. 1961, 1961, p17.
2 Ibid., p56.
3 Email correspondence with the author 4th March 2011.
4 http://www.skyatnightmagazine.com/news/prof-john-zarnecki-his-work-huygens-probe [10/03/2011].
5 Foreign Office Papers 371-159606, Memo from Foreign Office to Buckingham Palace, 12th July 1961.
6 Ibid.
7 Daily Worker 15th July 1961.
8 Guardian 15th July 1961.
9 http://www.bbc.co.uk/news/mobile/science-environment-12875848 [12/04/2011].
10 Guardian 6th April 2011.
11 Daily Worker 15th July 1961.
12 Ibid.
13 Daily Worker 15th July 1961.
14 Burrows, R.A.: Foreign Office Papers 371-159607, Internal Foreign Office memo, 18th July 1961, National Records Archive.
15 http://www.bbc.co.uk/programmes/p00fwcbn [10/04/2011].
16 Daily Worker 17th July 1961.
17 Ibid.

8. A Smile that Changed the World?

1 In the foreword of Scott, D. & A. Leonov: Two Sides of the Moon, 2004.
2 Siddiqi, A.A.: Challenge to Apollo: The Soviet Union and the Space Race, 2000, p145.

3 Ibid., p150.
4 Chertok, B.E.: Rockets and People. Volume III: Hot Days of the Cold War, 2009, p79.
5 Times 27th September 1968.
6 Jenks, A., The Cosmonaut Who Couldn't Stop Smiling: The Life and Legend of Yuri Gagarin, forthcoming 2011.
7 The names of all the cosmonauts were kept secret until they were in space or completed their mission. Only 12 of the 20 cosmonauts selected made spaceflights. There were several instances where the Soviets published photos where images of individual cosmonauts who had left the space programme were erased. See Oberg, E.J.: Uncovering Soviet Disasters, 1988, p164.
8 Email to the author 6th January 2011.
9 Chertok, B.E. Volume 1: Rockets and People, p21.
10 This and following quotes are from an interview with the author on 19th January 2011.
11 Burgess, C. & F. French: Into that Silent Sea: Trailblazers of the Space Era, 1961-1965, 2007, p27.
12 NASA translation #22, First Man in Space, 1st May 1961.
13 Alyona (Elena Andriyanovna) was the daughter of Valentina Tereshkova, cosmonaut in Vostok 6, and Andrian Nikolayev, cosmonaut in Vostok 3. She was the first person to have parents who had both been in space.
14 Nikolayev, A. & N. Kopylov: "Yuri Gagarin's Last Day", p23.
15 Scott, D. & A. Leonov: Two Sides of the Moon, 2004, p219 recalls the call sign to be 741, but Burgess, C. & R. Hall: The First Soviet Cosmonaut Team, 2009, p274 say it was 625.
16 Scott, D. & A. Leonov: Two Sides of the Moon, 2004, p219.
17 Bizony, P. & J. Doran: Starman: The Truth behind the Legend of Yuri Gagarin, 1998, p219.
18 Soviets were overtly secret and did not pretend otherwise. The Americans claimed openness from the start, but Reg Turnill writes on page 40 in his book The Moon Landings, "Looking back, with knowledge of what went on, I have modified my views considerably".
19 Scott, D. & A. Leonov: Two Sides of the Moon, 2004, p222.
20 The state of war ended in July 1951 for many former Western allies (Australia, Canada, Italy, New Zealand, South Africa, UK, USA) and the Soviet Union in early 1955.
21 Bloom, W.: in Washington to Caccia, at the FO, PREM 11/3316, 22nd April 1961, National Records Archive.
22 Rennie: in Washington to Whitehall and the FO, PREM 11/3316, 1st June 1961, National Records Archive.
23 Mason, P: in the UK permanent delegation to NATO, FO, PREM 11/3316, 1st June1961, National Records Archive.
24 The Times 11th July 1961.
25 Burgess, C. & F. French: Into that Silent Sea: Trailblazers of the Space Era,

1961-1965, 2007, p3.
26 Peoples History Museum CP/Cent/PL/04/04.
27 Full text in appendix.
28 The Times July 17th 2011.
29 The Times, letters page, 19th July 1961.
30 Steel, C.: Prem 11-3543, 15th July 1961, National Records Archive.
31 Interview with the author, 27th March 2011.
32 Rochdale Observer 22nd June 1963.
33 Andrew Jenks discusses this in fascinating detail in The Cosmonaut Who Couldn't Stop Smiling: The Life and Legend of Yuri Gagarin, forthcoming July 2011.
34 Inspired by Prof. Jim Aulich from the Manchester Metropolitan University and Nick Forder at the Museum of Science and Industry, see Manchester Evening News 6th July 2001.
35 Manchester Evening News 6th July 2001.
36 Manchester hosted an exhibition at the Waterside Arts Centre organised by Richard Evans (http://gagarin50.co.uk/). A film called "First Orbit" from Chris Riley portrayed an inspired recreation of what Gagarin would have seen had he looked down at the Earth for the entire flight (www.firstorbit.org).
37 Lovell, B: Astronomer by Chance, 1991, p2.
38 http://astrotalkuk.org/2008/04/20/episode12journey-to-the-moon/

Printed in Great Britain
by Amazon